QUANTUM THEORY
AND THE
SCHISM IN PHYSICS

By the same author

THE LOGIC OF SCIENTIFIC DISCOVERY

THE OPEN SOCIETY AND ITS ENEMIES
 Volume 1 The Spell of Plato
 *Volume 2 The High Tide of Prophecy: Hegel, Marx,
 and the Aftermath*

THE POVERTY OF HISTORICISM

CONJECTURES AND REFUTATIONS: THE GROWTH OF SCIENTIFIC
 KNOWLEDGE

OBJECTIVE KNOWLEDGE: AN EVOLUTIONARY APPROACH

UNENDED QUEST

THE SELF AND ITS BRAIN *(with J. C. Eccles)*

DIE BEIDEN GRUNDPROBLEME DER ERKENNTNISTHEORIE

REALISM AND THE AIM OF SCIENCE*

THE OPEN UNIVERSE: AN ARGUMENT FOR INDETERMINISM*

QUANTUM THEORY AND THE SCHISM IN PHYSICS*

 * *These last three volumes, taken together, constitute the*
 POSTSCRIPT TO THE LOGIC OF
 SCIENTIFIC DISCOVERY

QUANTUM THEORY
AND THE
SCHISM IN PHYSICS

KARL R. POPPER

From the POSTSCRIPT TO THE LOGIC OF
SCIENTIFIC DISCOVERY
Edited by W. W. Bartley, III

HUTCHINSON

London Melbourne Sydney Auckland Johannesburg

Hutchinson & Co. (Publishers) Ltd

An imprint of the Hutchinson Publishing Group

17–21 Conway Street, London W1P 6JD

Hutchinson Group (Australia) Pty Ltd
30–32 Cremorne Street, Richmond South, Victoria 3121
PO Box 151, Broadway, New South Wales 2007

Hutchinson Group (NZ) Ltd
32–34 View Road, PO Box 40–086, Glenfield, Auckland 10

Hutchinson Group (SA) (Pty) Ltd
PO Box 337, Bergvlei 2012, South Africa

First published 1982

Printed in the United States of America

British Library Cataloguing in Publication Data
Popper, Karl R.
 Quantum theory and the schism in physics.
 1. Quantum theory 2. Physics—Philosophical
 perspectives
 I. Title
 530.1'2 QC174.I3

ISBN 0 09 146170 7 cased
 0 09 146171 5 paper

TO JOHN ECCLES
and
TO PETER MEDAWAR

CONTENTS

Note on Numbering of Sections. The sections in each of the three
volumes of the *Postscript* are numbered consecutively, beginning with
section 1. The original section numbers, indicating the order of the sections
within the *Postscript* as a whole, are given in starred brackets in the Tables
of Contents. Ed.

EDITOR'S FOREWORD

1. *The* Postscript *and Its History.*

THIS book, *Quantum Theory and the Schism in Physics,* is the third volume of Sir Karl Popper's long-awaited *Postscript* to *The Logic of Scientific Discovery.* Although it was written some twenty-five years ago, it has never before been published. Yet it remains as timely as when it was first written: it challenges some of the most fundamental assumptions of current research in physics; and it also sketches a new cosmology.

Together with other parts of the *Postscript* (all of which are now being published), this volume was written mainly during the years 1951–56, at the time when *Logik der Forschung,* Popper's first published book (1934), was being translated into English as *The Logic of Scientific Discovery.*

The different volumes of the *Postscript* were originally part of a series of Appendices to *The Logic of Scientific Discovery,* in which Popper proposed to correct, expand, and develop the ideas of his first book. Some of these Appendices were in fact included in *The Logic of Scientific Discovery* when it was published in 1959. But one group of the Appendices took on a life of its own, and gradually grew into a single, closely integrated work far exceeding the original *Logik der Forschung* in length. It was decided to publish this new work—called the *Postscript: After Twenty Years*—as a sequel or companion volume to *The Logic of Scientific Discovery.* And it was accordingly set up in type, in galley proofs, in 1956–57.

Within a few months of the anticipated publication, however, the project came grinding to a halt. In *Unended Quest,* his intellectual autobiography, Sir Karl has reported of these galley proofs: 'Proof-reading turned into a nightmare. . . . I then had to have operations

on both eyes. After this I could not start proofreading again for some time, and as a result the *Postscript* is still unpublished.'

I remember this time vividly myself: I went to Vienna to visit Popper in the hospital there shortly after his operation for several detachments of both retinas; and we worked on the *Postscript* as he was recuperating. For a long time he could barely see, and we were very much afraid that he would become blind.

When he was able to see again, a great deal of work was done on the *Postscript*: several sections were added to it, and thousands of corrections were made to the galleys. But the pressure of other work had now become too great; and virtually nothing was added to the work after 1962. During the next, highly productive decade, after publishing *Conjectures and Refutations* (1963), Popper completed and published three new books: *Objective Knowledge: An Evolutionary Approach* (1972), *Unended Quest* (1974 and 1976), and (with Sir John Eccles) *The Self and Its Brain* (1977), as well as many papers. These were the years, and the works, in which his now famous theory of objective mind (and of Worlds 1, 2, and 3) was developed, and in which his approach was extended into the biological sciences.

Meanwhile, the *Postscript*, which represented the culmination of Sir Karl's work in the philosophy of physics, went unpublished. But not unread: most of Popper's closest students and colleagues have studied this work, and several have had copies of the galley proofs over the years. It is a source of great satisfaction to those like myself, who have known this book and been deeply influenced by it, to see it finally completed and shared with the general public.

The text that has now been edited for publication is essentially that which existed in 1962. Except in a few places, as marked, no major alterations have been made. It was felt that this was the appropriate approach to a work that had now acquired, through its influence on Popper's students and colleagues, an historical character—some twenty-five years having passed since its composition, and forty-five years since the writing of the original *Logik der Forschung*. Obviously many points would have been put differently today. But a complete revision by the author would have delayed publication indefinitely.

The editing has included bringing together the different versions

of some parts of the text, as they had accumulated over the years; copy-editing the book; and adding bibliographical and other notes for the reader's assistance. A few new additions made by Popper himself are clearly marked: they are presented in brackets and marked with a star: *. My own brief editorial and bibliographical notes are also in brackets, followed by the abbreviation 'Ed.'. Here I have in general followed the practice established by Troels Eggers Hansen, the editor of Popper's *Die beiden Grundprobleme der Erkenntnistheorie* (written in 1930-32 and published in 1979). Popper has been able to check the editorial work at a series of meetings which we have held at various places over the last two years—in Heidelberg, Guelph, Toronto, Washington, D.C., Schloss Kronberg, and at his home in Buckinghamshire. He has also added new prefaces to all of the volumes, and a new afterword to the second volume.

One major alteration in presentation has been made, at my own suggestion. To publish this large work under one cover would have been possible, but would have meant a heavy and unwieldy book beyond the means of many students of philosophy. Parts of the *Postscript*—including *Quantum Theory and the Schism in Physics*—will be of wide interest, of concern not only to philosophers and students of philosophy, but also to those many members of the general public who are interested in contemporary science and its interpretation, and in cosmology.

These parts of the *Postscript* are also, on the whole, independent of one another. This led me to suggest that the work be published in three separate volumes, in matching format, the whole constituting the *Postscript*. After some hesitation, Sir Karl agreed with this proposal, and also with the titles which I suggested for the three volumes.

Thus the *Postscript* is being published as follows:

Realism and the Aim of Science (Volume I)
The Open Universe: An Argument for Indeterminism (Volume II)
Quantum Theory and the Schism in Physics (Volume III).

Although these volumes can be read separately, the reader should be aware that they build a connected argument. Each volume of the

Postscript attacks one or another of the subjectivist or idealist approaches to knowledge; each constructs one or more components of an objective, realist approach to knowledge.

Thus in the volume now called *Realism and the Aim of Science,* Popper pursues 'Inductivism', which he sees as the chief source of subjectivism and idealism, through four stages: logical, methodological, epistemological, and metaphysical. He develops his theory of falsifiability, and charts its effects in demarcating scientific, non-scientific, and pseudo-scientific views from one another. And he presents his theory of corroboration as a way to express rational preference for one theory over another without resorting either to the subjective 'certainties' or to the objective 'justification' of conventional philosophies. In that first volume, Popper also discusses his relationship to those historical figures in philosophy, such as Berkeley, Hume, Kant, Mach, and Russell, who have contributed importantly to the subjectivist tradition; and he gives detailed replies to contemporary philosophical and scientific critics. In this volume Popper also attacks the subjective interpretation of the probability calculus, an interpretation rooted in the supposition that probability measures a subjective state of insufficient knowledge. In *The Logic of Scientific Discovery,* Popper had championed the frequency interpretation as an objective interpretation of the probability calculus. Now he also criticizes the frequency interpretation; and in its place he presents in detail his own propensity interpretation—an interpretation which has, during the past twenty years, found many champions. These ideas and arguments are applied and developed in the remaining volumes.

In *The Open Universe: An Argument for Indeterminism,* Popper presents a critique of both 'scientific' and metaphysical forms of determinism, and argues that classical physics does not, contrary to common opinion, presuppose or imply determinism any more than quantum physics does. Yet he finds that metaphysical determinism continues to underlie the work of many contemporary quantum theorists, opponents of determinism included. Popper traces the continuing role played within physics by subjective interpretations of probability to these metaphysical deterministic presuppositions.

There is a deep connection between the arguments of the first and second volumes, in their mutual concern with the freedom, creativity, and rationality of man.

The first volume, in its consideration of justification and rationality, rebuts a subjectivist and sceptical claim about the limits of criticism—and therewith the limits of rationality. If such a limit existed, then serious argument would be futile; and the appearance of it would be illusory.

The second volume, in its treatment of determinism, champions the claim that our rationality is limited in respect to the prediction of the future growth of human knowledge. If such a limit did *not* exist, then serious argument would be futile, and the appearance of it would be illusory.

Popper thus argues that human reason is unlimited with regard to criticism yet limited with regard to its powers of prediction; and shows that both the lack of limitation and the limitation are, in their respective places, necessary for human rationality to exist at all.

2. *This Volume of the* Postscript.

The *Postscript* as a whole, and *Quantum Theory and the Schism in Physics* in particular, are profoundly cosmological. It is in this final volume that the basic theme of Karl Popper's philosophy—that something can come from nothing—obtains its basis in physics.

It is not surprising that such an idea should meet incomprehension and stark resistance, for it opposes the dominant watchwords of our philosophical tradition:[1]

> There is nothing new under the sun.
> Nothing can be created or destroyed.
> *Ex nihilo nihil fit.*

The chief ideas of Popper's philosophy, and the leading themes of all three volumes of the *Postscript*, all relate to the basic theme that something can come from nothing. Scientific theories introduce new forms into the universe and cannot be reduced to observations: there is no such thing as scientific induction. The future is not contained in the present or the past. There is indeterminism in physics; and there is indeterminism in history, *ipso facto,* and also because new scientific ideas affect history and thus the course of the

[1]See W. W. Bartley, III: 'The Philosophy of Karl Popper. Part II. Consciousness and Physics: Quantum Mechanics, Probability, Indeterminism, the Body-Mind Problem', *Philosophia* 7, 3–4, July 1978, pp. 675–716.

physical universe. There is genuine emergence in biology. Value cannot be reduced to fact. Mind cannot be reduced to matter. Descriptive and argumentative levels of language cannot be reduced to expressive and signal levels. Consciousness is the spearhead of evolution, and the products of consciousness are not determined. Thus the importance of 'the open society' in 'the open universe'.

It is remarkable that a philosopher for whom consciousness and its products are so important should have at the core of his philosophical work a denunciation of those attempts—made by the majority of the best physicists during the past fifty years—to introduce consciousness into the heart of physics.

No connoisseur of the ironies of intellectual history and learned debate can fail then to appreciate the rare treat that is in store for him in this volume, where Popper carries his attack right to the center of current debate in quantum physics. At a time when most serious physicists are inductivist, subjectivist, positivist, instrumentalist—and when they attempt to ground these positions in physics—Popper proposes an interpretation of physics, and indeed an entire cosmology, which is deductivist, realist, anti-positivist, anti-instrumentalist. One can hardly imagine a more severe way to test the ideas of such a philosopher as Popper than to set his life in the middle of such a generation of physicists. His need over the past half century to forge his ideas, to test and shape them, to meet the actual and possible objections of the most abstract theoreticians of the most esoteric of modern subjects, has helped give to Popper's ideas their lapidary clarity.

It also gives them both a scientific importance and a cultural relevance that is not typical of contemporary philosophical thinking.

The Editor wishes to express his gratitude to The American Council of Learned Societies and to The American Philosophical Society for their generous support of his editorial work on these volumes; and to Professors Donald T. Campbell and F. A. von Hayek for their advice and support. He also wishes to thank his secretary, Nancy Artis Sadoyama, for her devoted and unfailing assistance.

ACKNOWLEDGEMENTS

I wɪsʜ to take this opportunity to thank my colleague John W. N. Watkins for the great encouragement which his unflagging interest has been to me. He has read this volume in manuscript and in proof, and has made the most helpful suggestions for improvements. It was at his suggestion that I decided to publish this *Postscript* as a separate work rather than, as originally planned, as a series of Appendices to *The Logic of Scientific Discovery*. But even more important for the completion of the work than these suggestions was his interest in its ideas.

I also wish to thank the co-translators of *The Logic of Scientific Discovery*, Dr Julius Freed and Lan Freed, who read most of the volume in galley proof, and made a great number of suggestions for improving its style. [They both died many years before its publication. Ed.]

Joseph Agassi was, during the period that this book was written, first my research student and later my research assistant. I discussed almost every section with him in detail, very often with the result that, on his advice, I expanded a statement or two into a whole new section—or, in one case, into a whole new part. [It became Part 2 of *Realism and the Aim of Science*.] His co-operation was of the greatest value for me.

I also wish to thank the London School of Economics and Political Science, which made it possible for me to benefit from Dr Agassi's assistance, and the Center for Advanced Study in the Behavioral Sciences (Ford Foundation) in Stanford, California, for giving me the opportunity of working uninterruptedly, from October 1956 to July 1957, on the galley proofs of this book, and for making it possible for Dr Agassi to help me during this period.

Penn, Buckinghamshire, *1959*.

Professor W. W. Bartley, III, was my student, and later my colleague at the London School of Economics, from 1958–63, and he worked closely with me during 1960–62 on this book. In 1978 he kindly consented to act as Editor of the *Postscript*. I am grateful to him for his assistance and for undertaking this arduous task. I am indebted to him more than I can say.

It is also a pleasure to thank several other persons who have in the intervening years worked with me on this *Postscript,* in particular, Alan E. Musgrave, David Miller, Arne F. Petersen, Tom Settle, and Jeremy Shearmur. Of these, David Miller and Arne Petersen should be specially mentioned because of the immense amount of work which they both did at various periods before 1970.

The London School has continued through all these years to help me by appointing a research assistant. For the thirteen years since my retirement in 1969 it has done so with the help of a grant from the Nuffield Foundation, to whom I wish to express my thanks. Chiefly responsible for this arrangement were my friend and successor, Professor John Watkins; the late Sir Walter Adams, Director of the School; and the present Director, Professor Ralf Dahrendorf, to whose warm friendship and great interest in my work I am deeply indebted.

Had the *Postscript* been published in the 1950s, I should have dedicated it to Bertrand Russell: Professor Bartley has told me that a letter to this effect exists in the Russell Archives at McMaster University.

I may mention, finally, that this *Postscript* (together with the translation of *The Logic of Scientific Discovery*) seemed to me almost ready in 1954. It was then that I chose its original title, 'Postscript: After Twenty Years', with an allusion to the publication of *Logik der Forschung* in 1934.

PENN, BUCKINGHAMSHIRE, *1982.*

AUTHOR'S NOTE

THIS book has been in the making for well over thirty years, though often laid aside. It is on the philosophy of physics, and physics has changed during this period (as has the author). As a result, the book contains several distinct parts. This note is the author's attempt to guide the reader through them.

The main part, from p. 97 to the end, remains as it was in galley proofs in February 1957, with a few minor stylistic changes and a few added bibliographic references and footnotes.

What comes before this main part is called 'Introductory Comments'. It contains several essays written in 1980–81, as well as an essay written in 1966 (with an addition written in 1981). These are several attempts to introduce later developments of my views.

These developments are, partly, simplifications of old ideas (such as, especially, section IX of the 'Preface 1982'); partly they are reactions to—or more precisely against—the changing atmosphere in the orthodoxy of quantum mechanics, the so-called Copenhagen interpretation. One might distinguish the following periods:

1927–1932. The period of what I later called the 'end-of-the-road thesis': Quantum mechanics was to be regarded as the Last Revolution in physics, since it had reached the inherent limits of knowledge. For this was the real content of the indeterminacy relations. (My reply was: they are nothing of the kind—merely scatter relations.)

1932–1936. Some doubts came with the first new particles (which really meant the end of the period in which quantum mechanics could see itself as the culmination of the electromagnetic theory of matter). Einstein saw clearly the absurdity of the claim that quantum mechanics was the end—that it was 'complete'. This led to EPR (the paper of Einstein, Podolsky, and Rosen), and to Bohr's reply, which insists on its completeness.

1936–1948. The end-of-the-road thesis was abandoned by Pauli. Bohr persisted in his reply to EPR. Einstein proposed (in a short paper in *Dialectica*, 1948) the thesis that quantum mechanics implies action at a distance; but this thesis remained unnoticed.

1948–1964. New experimental discoveries (not quite matched by theoretical developments) swept away the end-of-the-road thesis. The EPR was reformulated by David Bohm, but otherwise not much remembered.

1964–1981. J. S. Bell reformulates the EPR experiment in the light of Einstein's discovery that quantum mechanics implies action at a distance. New experiments, especially Aspect's experiments, seem to confirm action at a distance.

My own reaction to this: I feel not quite convinced that the experiments are correctly interpreted; but if they are, we just have to accept action at a distance. I think (with J.-P. Vigier) that this would of course be very important, but I do not for a moment think that it would shake, or even touch, realism. Newton and Lorentz were realists and accepted action at a distance; and Aspect's would be the first crucial experiment between Lorentz's and Einstein's interpretation of the Lorentz transformations.

Realism is the message of this book. It is linked with objectivity, also in probability theory. This link yields the propensity interpretation. Realism is linked with rationalism, with the reality of the human mind, of human creativity, and of human suffering.

INTRODUCTORY COMMENTS

PREFACE 1982: ON A REALISTIC AND COMMONSENSE INTERPRETATION OF QUANTUM THEORY

I T is the great task of the natural sciences and of natural philosophy to paint a coherent and understandable picture of the Universe. All science is cosmology, and all civilizations of which we have knowledge have tried to understand the world in which we live, including ourselves, and our own knowledge, as part of that world.

In this endeavour of understanding the world, physical science, with its wonderful combination of speculative creativity and openness to experience, has become of fundamental importance. This was not always so, and it may not remain so for ever. But at present, it seems as if our world was a world of physics long before a world of chemistry emerged, and far longer before a world of biology emerged.

Today, physics is in a crisis. Physical theory is unbelievably successful; it constantly produces new problems, and it solves the old ones as well as the new ones. And part of the present crisis—the almost permanent revolution of its fundamental theories—is, in my opinion, a normal state of any mature science. But there is also another aspect of the present crisis: it is also a crisis of understanding.

This crisis of our understanding is roughly as old as the Copenhagen interpretation of quantum mechanics. It is thus a little older than the original edition of *The Logic of Scientific Discovery*. In this part of the *Postscript* I have tried to make again a number of proposals intended to clarify what underlies this crisis of understanding.

I

In my view, the crisis is, essentially, due to two things: (a) the intrusion of subjectivism into physics; and (b) the victory of the idea that quantum theory has reached complete and final truth.

1

Subjectivism in physics can be traced to several great mistakes. One is the positivism or idealism of Mach. It spread to the British Isles (where it had been originated by Berkeley) through Russell, and to Germany through the young Einstein (1905). This view was rejected by Einstein in his forties (1926), and it was deeply regretted by the mature Einstein (1950).[1] Another is the subjectivist interpretation of the calculus of probability, which is far older and which became a central dogma of the theory of probability through the work of Laplace.

The central issue here is *realism*. That is to say, the reality of the physical world we live in: the fact that this world exists independently of ourselves; that it existed before life existed, according to our best hypotheses; and that it will continue to exist, for all we know, long after we have all been swept away.

I have argued in favour of realism in various places.[2] My arguments are partly rational, partly *ad hominem*, and partly even ethical. It seems to me that the attack on realism, though intellectually interesting and important, is quite unacceptable, especially after two world wars and the real suffering—avoidable suffering—that was wantonly produced by them; and that any argument against realism which is based on modern atomic theory—on quantum mechanics—ought to be silenced by the memory of the reality of the events of Hiroshima and Nagasaki. (I say this full of admiration for modern atomic theory and quantum mechanics, and of the scientists who have worked and are now working in this field.)

Realism was first challenged by scepticism, especially by certain arguments of Descartes, who, in order to get the better of scepticism, tried to draw from it its most radical and, indeed, absurd conclusions. It was, next, challenged by Bishop Berkeley, who, like Cardinal Bellarmino, feared that science, if taken seriously, would supersede Christianity.[3]

Thus arose an idealistic (or even positivistic[4]) philosophy, a philosophy that takes our subjective experiences—especially our

[1]The dates are those of Einstein's first paper on special relativity, a visit of Heisenberg's to Einstein in Berlin, and my visit to Einstein in Princeton.

[2]See, especially, *Conjectures and Refutations*, section 6 of Chapter 3 (written in 1954, first published in 1956); and *Objective Knowledge*, Chapter 2, especially section 5.

[3]See *Conjectures and Refutations*, Chapter 3, sections 1 and 2.

[4]See *Conjectures and Refutations*, Chapter 6.

perceptions, our observations—as more secure, more certainly real, than the physical reality, which, positivism alleges, is merely our mental construction. However, one can look at these matters in a different way. One can look at our mental constructions as serving the task of helping us to survive in a real world which is still largely unknown to us.

We owe to Kant the first great attempt to combine a realistic interpretation of natural science with the insight that our scientific theories are not simply the result of a description of nature—of 'reading the book of nature' without 'prejudice'—but that they are, rather, the products of the human mind: 'Our intellect does not draw its laws from nature, but it imposes its laws upon nature.'[5] I have attempted to improve this excellent Kantian formulation as follows: 'Our intellect does not draw its laws from nature, but it tries—with varying success—to impose upon nature laws which it freely invents.'[6]

Thus theories are our own inventions, our own ideas: this has been clearly seen by the epistemological idealists. But some of these theories are so bold that they can clash with reality: they are the testable theories of science. And when they clash, then we know that there is a reality: something that can inform us that our ideas are mistaken. And this is why the realist is right.[7]

(Incidentally, this kind of information—the rejection of our theories by reality—is, in my view, the only information we can obtain from reality: all else is of our own making. This explains why our theories are all coloured by our human point of view, but less and less distorted by it as our search goes on.)

So much for realism—for 'scientific realism', for 'critical realism'.

The mistaken view that 'scientific realism' implies that our scientific theories are to be based on what we can actually observe, that is, on the information, the 'data', given to us by reality—a view clearly rejected by Einstein in 1933,[8] though it is still popular even among theoretical physicists—has led, as often before, to the

[5]Kant: *Prolegomena*, end of section 37.
[6]*Conjectures and Refutations*, Chapter 8, p. 191.
[7]Compare with this paragraph *Conjectures and Refutations*, p. 117, second new paragraph (Chapter 3, section 6).
[8]See Einstein: *On the Method of Theoretical Physics* (The Herbert Spencer Lecture, 1933), Oxford, 1933.

intrusion of a subjectivist or positivist or idealist or solipsist interpretation of science.

II

A second source of the intrusion of subjectivism is connected with the rise of *probabilistic physics,* which first became of fundamental importance for the theory of matter with Maxwell and Boltzmann, who of course had many distinguished predecessors.

This new probabilistic physics was regarded for a long time as connected with our *lack of knowledge.* [9] Even during the 1930s, and perhaps longer, it was still thought that probabilistic considerations enter physics only because we cannot possibly know the precise positions and momenta of all molecules in a gas. This forces us to attribute probabilities to the various possibilities; a method that is the basis of statistical mechanics. If we could know, if we could be certain, of all the positions and momenta of the particles in question, we should not have to fall back on probabilities.

Thus a link was forged between lack of knowledge on the one side and probabilistic or statistical physics on the other.

The assumption of this link was pretty universal among scientists up to, say, 1939, and even afterwards. Einstein for example defended it against my *L.Sc.D.*, where I had criticized it (but he gave up the defence during a discussion in 1950). [10]

It can be shown that this view of the role played by our subjective lack of knowledge, or our uncertainty, was paramount at least in Heisenberg's early interpretations of those formulae which he called the 'uncertainty relations' and which, he thought, *explained* why quantum mechanics had a statistical character. (Compare with this *L.Sc.D.*, especially section 75.)

[9] See Bryce S. DeWitt: 'Quantum mechanics and reality', in Bryce S. DeWitt and Neill Graham: *The Many-Worlds Interpretation of Quantum Mechanics* (Princeton, 1973), pp. 160–61.

[10] I used two arguments which convinced him very quickly. The one was that even if we would know everything, we should still have to derive *statistical information* from this knowledge in order to answer what are essentially *statistical* problems, such as explaining gas pressure or the intensity of spectral lines. (This can also be seen from the way in which we have to derive statistical information from questionnaires.) The second, closely related, argument was the more general logical argument that we need statistical or probabilistic premises (a statistical or probabilistic theory) to derive statistical conclusions.

I think that there can be little doubt that, as a matter of history, these two points explain the intrusion of subjectivism into physics. Together, they led to a positivist or idealist or subjectivist rejection of realism, motivated by the belief that a *fundamentally and irreducibly statistical physics has to be explained by a fundamental and irreducible barrier to our (subjective) knowledge*—a barrier (the uncertainty relations) which in its turn is objective, to be sure, but still a barrier to what a subject can know.

It must be admitted that the belief in the objectivity of this barrier led, in time, to a change: the part played by probability was now seen differently. Thus quantum physics became regarded as objectively indeterministic, and probability as something objective (as I had argued in *L.Sc.D.*).

But my historical conjecture is that by that time, the subjectivist dogma was too deeply entrenched within the ruling interpretation of quantum mechanics, the so-called Copenhagen interpretation; and even Heisenberg's occasional talk of objective possibilities (by which he meant something very similar to my propensities) did not remove the subject—the observer—from the Copenhagen interpretation; nor was it intended to do so.

I have sketched this story here briefly because it explains, in part, how the great quantum muddle started: why Heisenberg's so-called 'uncertainty relations' were, at least for a long time, interpreted as limits to our subjective knowledge (rather than as those objective statistical 'scatter relations' which I suggested to replace Heisenberg's interpretation in *L.Sc.D.*, as long ago as 1934); and why, when their objective side came to the fore, they were still interpreted as statements about the impossibility of certain measurements *due to the non-existence of the entities measured*—instead of as statements about the impossibility of producing scatter-free (dispersion-free) quantum states.

And the whole terminology, introduced in the early period of the theory, conspired to make the muddle worse and worse.

III

Another source of the present crisis in physics is the persistence of the belief that quantum mechanics is final and complete. And the

strongest reason for my own opposition to the Copenhagen interpretation lies in its claim to finality and completeness.

The Copenhagen interpretation—or, more precisely, the view of the status of quantum mechanics which Bohr and Heisenberg defended—was, quite simply, that quantum mechanics was the last, the final, the never-to-be-surpassed revolution in physics; together with the thesis that the truth *about* the situation in physics can be established by arguments based upon physics itself: more precisely, upon Heisenberg's relations of uncertainty or indeterminacy. These were claimed to show that *physics has reached the end of the road*; that a further breakthrough is no longer possible, although, of course, much is still to be done by way of elaboration and application of the new quantum mechanics: in other words, by way of 'normal science', as distinct from revolutionary science.

I hardly need to say that in 1930, as today, I regarded scientific revolution as typical of *all* great science. I immensely admired Bohr and Heisenberg as revolutionary scientists. But this epistemological claim I regarded, and still regard, as outrageous.

It is a claim that other people regarded as outrageous too. It is also a claim that may have been forgotten by now.

I think that it has indeed been forgotten: nobody mentions it any longer, even though it is central to the present discussion of quantum theory (especially since J. S. Bell opened the way to its experimental investigation).

Let us very briefly sketch the history.

Einstein and others who appreciated quantum mechanics as a revolutionary breakthrough but did not accept its finality, or what I propose to call the 'end-of-the-road thesis', believed that there must be a further, deeper level in physics, a level beyond quantum mechanics.

Einstein supported this view for a long time by the argument (in my opinion mistaken) that quantum mechanics is a probabilistic theory and that probability comes into physics only because of our lack of knowledge (see the previous section).

I always regarded this subjectivist view of probability as mistaken, and I think that Einstein gave it up—perhaps finally—during our discussion in 1950. But even those who did not agree with Einstein about this particular reason for rejecting the end-of-the-road thesis did agree that there may be a deeper layer of physical

reality beyond that described by the equations of quantum mechanics—perhaps in nuclear physics. After all, quantum mechanics had, essentially, developed as the theory of the electronic shells surrounding the atomic nucleus.

But this was certainly not Heisenberg's view. I spent an evening with him (I think in 1935) when he came to Vienna. At that time he suggested that quantum mechanics would *not* be deepened by the investigation of the nucleus. He predicted, rather, that the situation prevailing in the atomic nucleus may make the indeterminacy worse: it may turn out that the limits to our knowledge in nuclear theory may be even more severe than those in atomic theory (the theory of the electronic shell), and that we shall be prevented forever from understanding the working and the stability of the nucleus even to the extent to which we can understand the working and stability of the electronic shells.

Today it is fairly obvious that those who disbelieved the end-of-the-road thesis were right. Heisenberg himself contributed to transcending it. In fact, the thesis looks so absurd by now that I suppose few physicists will believe that it was ever held; or, if it was held, that it was taken seriously.

Yet this end-of-the-road thesis did underlie the great clash of the titans, the discussions between Albert Einstein and Niels Bohr.

It is commonly accepted that Einstein was defeated in this fight. (See also below, the 'Introduction'.) But the truth is quite different. The real issue between Einstein and Bohr was what both called, somewhat vaguely and misleadingly, *the problem whether or not quantum mechanics was complete.*

The term 'complete' has been used in several senses during this discussion; but fundamentally, and at the beginning of the discussion, it certainly was intended to help formulate *the problem whether or not quantum mechanics was (at least in principle) the end of the road in physics.*

It is important to remember in this connection that Einstein never regarded any of his own revolutions as a *final* breakthrough. His own photon theory, and the need to use it together with the wave theory of light, which really established what was later called wave-particle duality, he regarded as a stop-gap, although it brought him almost to the threshold of the theory of matter waves; that is to say, to the extension of particle-wave duality from the

theory of light to the theory of matter. His special theory of relativity he also regarded, rightly, as unsatisfactory (for several reasons, especially because it merely replaced absolute space by the absolute set of inertial systems). Of his general theory he said that it was 'ephemeral'; and from the moment of its conception to the end of his life he tried to supersede it.

It was different with Heisenberg and Bohr. Heisenberg's own breakthrough, which was accompanied by an experience of the intuitive vision of the new theory, made a tremendous impression on him. As he puts it himself, 'the most important criterion of truth' is the intuitive vision of the simplicity of the solution which 'at the very end' illuminates everything.[11]

I think that this great experience, the vision of an 'illuminating simplicity', was decisive for Heisenberg's attitude. It made him feel 'this is *it*': this was the end of the road, the final truth. And it made him blame others—especially Einstein—for not seeing that it was indeed the end of the road in physics. He used the German word 'endgültig'. But the English translation, 'finally valid', fails to convey the feeling of having reached the final solution that is conveyed by the German term 'endgültig'. It was in trying to convey this feeling that I hit on the phrase 'end of the road'.

Thus in his Preface to the Einstein-Bohr correspondence (written probably in 1968), Heisenberg tries to describe the situation which separated Einstein from Heisenberg himself as well as from his teacher Max Born. He says that 'Einstein agreed with Born that the . . . mathematical formalism of quantum mechanics correctly represents the phenomena of the [electronic] shells of the atom'. But, he complains that 'Einstein did not want to acknowledge that quantum mechanics represented a finally valid, and even less a complete, description of these phenomena'.[12] Here 'finally valid' (endgültig) should, I think, be translated by 'ultimately valid'—indeed, by 'the end of the road'.

When he wrote this, Heisenberg hardly recalled that Einstein's attitude towards his own theories had been far more critical than his attitude towards those of Heisenberg.

A similar remark about the conflict can be found in the memorial

[11]Werner Heisenberg, *Der Teil und das Ganze*, p. 138.

[12]See *Albert Einstein—Hedwig und Max Born: Briefwechsel: 1916–1955*, Munich, 1969, or *The Born-Einstein Letters*, New York, 1971, pp. ix–x.

volume *Werner Heisenberg*.[13] Here C. F. von Weizsäcker, a close friend and collaborator of Heisenberg's, writes (p. 66) that Einstein thought that quantum mechanics is not yet 'die endgültige Physik' (the finally valid physics, the end of the road in physical science). Einstein's critical attitude, thus described, is taken as a sign of his not quite sufficient understanding, as a sign of his old-fashioned backwardness, which led to his loss of contact with quantum mechanics.

Before leaving Heisenberg and his very understandable attitude, I may perhaps just mention that his views changed a great deal. It was he who led a generation of physicists to accept the absurd view that one can learn from quantum mechanics that 'objective reality has evaporated'.[14]

The case of Niels Bohr is somewhat different.

When, thanks to Victor Weisskopf, I first had the great opportunity, in 1936, of talking to Bohr, he impressed me as the most wonderful person I had ever met or would be likely to meet. He was everything a great and good man could be. And he was irresistible. I felt that I must be wrong about quantum mechanics, even though I certainly could not say that I now understood it, rationally, in Bohr's sense: I did not. But I was overwhelmed.

Bohr's attitude, I think, was very different from Heisenberg's. I think that Bohr was, basically, a realist. But quantum theory had been, from the very start, a riddle for him: nobody could have been more keenly aware of the depth of the difficulties that beset his atom model of 1913. He never got rid of these difficulties. In all his marvellous successes there also was always embedded some failure: something incomprehensible, impenetrable, prevented him from reaching clarity. He had perhaps never experienced that satisfaction that Heisenberg had experienced, that blinding flash of light that illuminated everything. When he accepted quantum mechanics as the end of the road, it was partly in despair: only classical physics was understandable, was a description of reality. Quantum mechanics was not a description of reality. Such a description was

[13]Carl Friedrich von Weizsäcker and Bartel Leendert van der Waerden, *Werner Heisenberg*, Munich, 1977. See also C. F. von Weizsäcker, 'Gedenkworte für Werner Heisenberg', in *Orden Pour le Mérite für Wissenschaften und Künste: Reden und Gedenkworte*, Heidelberg, 1976, pp. 63–81.

[14]*Daedalus* **87**, 1958, pp. 95–108.

impossible to achieve in the atomic region; apparently because no such reality existed: the understandable reality ended where classical physics ended. The nearest to an understanding of atoms was his own principle of complementarity.

This principle told us something about the limits of classical physics and thus of understanding. We could understand classical particles and we could understand classical waves and we could understand that the particle description and the wave description were incompatible. They were incompatible, yet both necessary. That was the limit to which our understanding could penetrate. It was the end of the road—but in a sense quite different from Heisenberg's, even though both agreed that an act of 'renunciation' of our understanding was needed.

Bohr's link of the 'principle of complementarity' with the 'dualism of particle and wave' broke down when Born's interpretation of the square of the wave amplitude as a probability for finding the particle was accepted. For this meant, indeed, the acceptance of the particle interpretation as basic. (On the relation between the wave-particle duality and Heisenberg's relations, see note 38a to the 'Introduction' below.) From that moment chaos ruled in the Copenhagen camp. (For example, a great physicist and mathematician like Pascual Jordan could make assertions such as those discussed in *L.Sc.D.*, pp. 454–55.) Murray Gell-Mann, one of the very few physicists who dares to criticize Bohr, puts it perhaps a little too bluntly when he says that 'Niels Bohr brainwashed a whole generation of theorists into thinking that the job' (that is, an adequate philosophical presentation of quantum mechanics) 'was done 50 years ago'.[15] But I rather think that most of the physicists, theorists included, did not care too much, and simply relied on Bohr in a matter in which they did not feel competent.

Bohr also supported the end-of-the-road thesis, largely forgotten now, which was the very center of his great struggle with Einstein. That it was the issue, at least in those days, one can see very clearly from the famous paper by Einstein, Podolsky, and Rosen (1935).[16]

[15]M. Gell-Mann in Douglas Huff and Omer Prewett, *The Nature of the Physical Universe: 1976 Nobel Conference*, New York, 1979, p. 29.

[16]A. Einstein, B. Podolsky, and N. Rosen, 'Can Quantum-Mechanical Description of Physical Reality Be Considered Complete?', *Physical Review* 47, 1935, pp. 777–780.

IV

But before this, already in 1932, John von Neumann's *Die mathematischen Grundlagen der Quantenmechanik* came out—a very, very difficult book (for me, at least).[17] In this book von Neumann gave a mathematical proof purporting to establish once and for all the final, the end-of-the-road character of quantum mechanics: he proved that all those were mistaken who, like Einstein, thought that there may be a layer of physical reality deeper than the one represented by quantum mechanics.

In order to make this proof quite general, von Neumann introduced a concept that became very famous: the concept of 'hidden variables'. A 'hidden variable' was anything to be taken into account in atomic theory (in the sense in which atomic theory included the nucleus as a matter of course) that was not yet taken into account by quantum mechanics. Von Neumann proved (or so we were told) that such hidden variables could not exist in quantum mechanics; or, according to a slightly different interpretation, he proved that the existence of hidden variables contradicted quantum mechanics.

Now it so happened that in the same year in which von Neumann's book was published, two new particles were discovered: the neutron and the positron.

Were these not (previously) hidden variables? And if not, what would have been?

Quantum mechanics was, after all, not an abstract physical formalism but a theory of something very concrete: a theory of atoms, a theory of their structure as possessing a positively charged nucleus and a shell structure of negative electrons that explained, in principle, very concrete properties of the chemical elements.

In 1931, these structures were still assumed to be structures of only *two* material particles—the electron and the proton; in addition, of course, to a non-material particle, the photon. All atomic structures were to be explained in these terms; and the nuclei were composed of them, in an unknown (and perhaps unknowable) manner.

Then, during 1932, the positron and the neutron were discovered; and in 1933, Pauli announced at the Solvay conference in Brussels

[17]As to von Neumann's proof of the impossibility of hidden variables, I could not find any convincing argument in it beyond the exclusion of dispersion-free effects.

his bold idea of the existence of another particle, for which Fermi suggested the name 'neutrino'.

It is of some interest to discuss briefly the reasons which prevented the leading physicists from realizing that these particles refuted their claim that quantum mechanics, as for instance formulated by von Neumann, had been the end of the road in physics (endgültig) and complete (vollständig).

(a) The positron. It has been reliably claimed that Anderson's discovery of the positron was at first most unwillingly received by Bohr, Heisenberg, Schrödinger, and also by Eddington. However, when it became clear that the discovery could not be rejected, it was, rightly I think, hailed as a success for quantum mechanics, in the form given to it by Dirac. For Dirac had indeed obtained from his theory the prediction of the existence of a particle with a positive charge. He had supposed that this would refer to the proton; but it fitted the positron far better. So there was some fairly good reason to say that the positron, far from being a refutation, was a confirmation of quantum mechanics.[18]

(b) Chadwick's neutron was perhaps less of a problem: it could be, and it was, interpreted at first as composed of a proton and an electron. It turned out, however, that there was a serious difficulty here: the theory—quantum mechanics—did not succeed in explaining this composition. So it was, in time, accepted as a new particle, which may arise either through the common transition of a proton and an electron into a neutron, or by the emission of a positron from a proton. (It is interesting to read in a recent paper by Ruggero Maria Santilli that, in his theory, this 'first structure model of the neutron' is being revived, by 'resolving the technical difficulties which [had] led, historically, to the abandonment of the model'.[19]

[18]This is not of course anything like the whole story. For a fuller account see Norwood Russell Hanson, *The Concept of the Positron*, Cambridge, 1963, especially pp. 143–158. These pages ought to be read by everybody interested in this story.

[19]Ruggero Maria Santilli: 'An Intriguing Legacy of Einstein, Fermi, Jordan, and Others: The Possible Invalidation of Quark Conjectures', *Foundations of Physics* 11, 5/6, 1981, pp. 383–472, especially 448. Santilli refers here to his paper in *Hadronic Journal* 2, 1979, p. 1460, section 2.4. See also Santilli's book *Foundations of Theoretical Mechanics I: The Inverse Problem in Newtonian Mechanics*, New York, 1978.

These difficulties, he says, 'were *all* due to the assumption that atomic mechanics' [Santilli's name for quantum mechanics] 'applied within the neutron, and [they] are removed when a generalized mechanics is used'.

(c) It is then understandable that, in this way, the end-of-the-road people (to whom Pauli at that time belonged) were no longer shocked by the neutrino (which had not yet been discovered: it was a purely speculative particle), even though it should have been recognized as a further 'hidden variable'. But by now they obviously had unconsciously shifted their positions, including their ideas about hidden variables. Originally, and especially in their early discussions with Einstein, the name 'quantum mechanics' was used for the theory of the atom (including the nucleus *as a matter of course*). Now the name was more and more used simply to refer to the powerful mathematical formalism, due mainly to Heisenberg, Born, Jordan, Schrödinger, and Dirac, Gordon, and Klein; a formalism which, although at first a theory of the motion of the (negative) electron in the positive field of the nucleus, was more and more widely applied. But this formalism, of course, also changed very considerably, and grew far beyond the original theory of Heisenberg, Born, and Jordan, the theory which was to be the end of the road in physics.

(d) When, in 1935, Yukawa predicted a new particle, later called the meson, the search for this particle led to the discovery of several different mesons. But none of these, or of the many other elementary particles, were identified as (formerly) 'hidden variables'; even after 'hidden variables' were freed from von Neumann's original postulate that they had to turn the probabilistic theory into a deterministic one—a postulate which some people (among them also myself) have regarded as impossible to satisfy.

Meanwhile quantum mechanics has grown. Its name was, more or less, preserved, since the theory preserved a number of characteristic principles (the commutation relations, Heisenberg's uncertainty relations, Pauli's exclusion principle). Thus the name 'quantum mechanics' was applied to quantum electrodynamics and quantum field theory (and even to quantum chromodynamics, and to theories which introduced new quantum numbers and, therefore, previously 'hidden' variables). And Bohr's defence, against Einstein, of the thesis that quantum mechanics was 'complete' (in the

13

end-of-the-road sense) went on being regarded as valid, after the theory had changed radically and after its incompleteness had become too obvious to be regarded any longer as a possible issue of that great struggle.

Admittedly, the theory had not changed in precisely the way Einstein had hoped. But this was, decidedly, not the issue: Einstein was never dogmatic even about his best supported theories, and certainly not about his hopes.

Yet dogmatism was imputed to him by those who, like Heisenberg (in the passage quoted above from his preface to the Einstein-Bohr correspondence), stuck to the end-of-the-road thesis after all these new and perhaps unwelcome revolutions had proved that the end was still far off.

I have mentioned Santilli, and I should like to say that he—one who belongs to a new generation—seems to me to move on a different path. Far be it from me to belittle the giants who founded quantum mechanics under the leadership of Planck, Einstein, Bohr, Born, Heisenberg, de Broglie, Schrödinger, and Dirac. Santilli too makes it very clear how greatly he appreciates the work of these men. But in his approach he distinguishes the region of the 'arena of incontrovertible applicability' of quantum mechanics (he calls it 'atomic mechanics') from *nuclear mechanics* and *hadronics*, and his most fascinating arguments in support of the view that quantum mechanics should not, without new tests, be regarded as valid in nuclear and hadronic mechanics, seem to me to augur a return to sanity: to that realism and objectivism for which Einstein stood, and which had been abandoned by those two very great physicists, Heisenberg and Bohr.

V

The present situation in physics can only be understood against this background. There are elements in the present situation other than those referred to above, for example in the work of J. S. Bell (which I shall discuss below). But even Bell's work is directly connected with these problems of realism, probability, and completeness or finality through the argument of Einstein, Podolsky,

and Rosen; an argument that was, in its original form, directed against the opposition to realism characteristic of the Copenhagen interpretation.

During the past decade there have in fact been some new and exciting developments in quantum theory. These relate to the famous thought experiment (1935) of Einstein, Podolsky, and Rosen (to which I shall refer as EPR). This was an experiment that Einstein produced very shortly after I tried to produce a similar experiment. My own experiment, which I published in *Naturwissenschaften* and also in *Logik der Forschung* (1934), was invalid.[20] Because of my own involvement with a similar—although, unfortunately, mistaken—argument, I think it fair to say that I was one of the first philosophers to realize the significance of EPR, and I have always been very interested in it. Among physicists it created quite a stir right from the start; but the philosophers ignored it.

It is the EPR argument which has led to these new developments

[20][K. R. Popper: 'Zur Kritik der Ungenauigkeitsrelationen', *Naturwissenschaften* **22**, 48, November 30, 1934, pp. 807–8; and *The Logic of Scientific Discovery*, section 77 and note on p. 236 and notes *3 and *4 to section 77; see also *L.Sc.D.*, Appendices *xii and *xi, section 3; see also Chapter I, section 17, of the present volume. Max Jammer has written, in *The Philosophy of Quantum Mechanics*, 1974, pp. 174 and 178n, as follows: Popper's 'scientific début—while he was still earning his living as an elementary school teacher—was a paper in *Die Naturwissenschaften* (1934) which contained, as he later called it, "a gross mistake for which I have been deeply sorry and ashamed ever since". It is not impossible that it was precisely this "mistake" which prompted Einstein (who immediately recognized the error) to publish, together with Podolsky and Rosen, the argument against the incompleteness of quantum mechanics. . . . According to Professor Nathan Rosen, it might have been possible that Popper influenced Einstein (private communication, April 22, 1967); but according to Mrs. Polly Podolsky, who kept in close touch with her (in 1966 deceased) husband's work, it was unlikely that Popper's work could have reached Einstein before the first draft of the Einstein-Podolsky-Rosen paper was written (letter from Mrs. Polly Podolsky, dated August 1, 1967). Sir Karl Popper expressed his opinion on this issue in a letter dated April 13, 1967, as follows: "Now about your 'hunch'. This I find extremely flattering; but I must say that it never occurred to me before reading your letter: the possibility that a *gross mistake* made by a nobody (like myself) may have had any influence on a man like Einstein never entered my head. From a purely temporal view your hunch cannot be entirely ruled out . . ." (letter to author). That from a purely logical view his reasoning was closely related to the Einstein-Podolsky-Rosen argument was clearly recognized by Popper when he wrote in a footnote to p. 244 of Ref. 3–5 (1959): "Einstein, Podolsky, and Rosen use a *weaker* but *valid* argument." In fact, one may say that Popper's is an overdetermined Einstein-Podolsky-Rosen argument.' See Jammer's discussion of several other lines of influence in 'The Prehistory of the EPR Argument', in section 6.2 of his work, pp. 166–181. Ed.]

15

in quantum theory, in particular to what is called the 'Bell inequalities', and the problem of locality.[21]

Two stages of the discussion have to be distinguished in order to understand the development which has led up to the present state of affairs. The first is the original EPR paper itself, which is written *against* action at a distance. It made clear for the first time that the Copenhagen interpretation of quantum theory implied action at a distance—even for very great distances.

Then there is another stage, to be credited to David Bohm.[22] This stage differs in various ways from the first. Bohm's argument was based on polarization or *spin*. Spin plays no part in the original EPR, which was concerned with position and momentum, but spin does play a crucial role in the version due to Bohm. The original EPR was a thought-argument; it seemed at first impossible to realize it experimentally. But the second form, due to Bohm, has turned out to be experimentally testable.

VI

The original EPR argument was, essentially, a challenge to Heisenberg's interpretation of his so-called 'uncertainty relations' or 'indeterminacy relations'; that is to say, of his formulae

$$\Delta p_x \, \Delta q_x \gtrsim h / 2\pi \tag{1}$$

These formulae state that two intervals or ranges—here the interval or range Δp_x of the momentum in the direction of the x-axis and the interval or range Δq_x of the position along the x-axis—are so related, that their product cannot become smaller than Planck's constant h divided by 2π; which means, of course, that the smaller we make Δp_x the larger becomes Δq_x, and *vice versa*.

A formula like this is characteristic of any wave theory. It can be

[21]See my paper, 'Particle Annihilation and the Argument of Einstein, Podolsky, and Rosen', in Wolfgang Yourgrau and Alwyn van der Merwe, eds.: *Perspectives in Quantum Theory*, 1971, pp. 182–98. I am no longer satisfied with this paper. See Bell's excellent reply to it in *Science* 177, 1972, p. 880, and also Bell's contribution to B. d'Espagnat, ed.: *Foundations of Quantum Mechanics, Proceedings of the International School of Physics "Enrico Fermi"*, 1971, p. 171. See also the discussion in John F. Clauser and Michael A. Horne: 'Experimental Consequences of Objective Local Theories', *Physical Review* 10, 2, 15 July 1974, pp. 526–535.

[22]David Bohm: *Quantum Theory*, 1951; and *Physical Review* 85, pp. 169–193.

derived, for example, from applying a wave theory of light to a situation where a beam of light moving in the x-direction falls on a screen with a narrow slit in it.

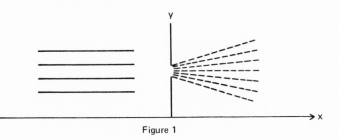

Figure 1

$$\Delta p_y \; \Delta q_y \geqq h$$

The narrower the slit Δq_y the greater will be the angle of the scatter of the momenta of the beam in the y-direction, after the light has gone through the slit. (A similar result is obtained from Schrödinger's wave theory for electrons.)

The formula (1) can be obtained, for example, by applying the wave equation to the physical situation (the initial conditions) described by this example. This shows that our formulae (1) needs to be *interpreted:* it is obtained from the theory when we apply the theory to such cases as a beam falling on a screen with a narrow slit, or to similar cases.

Now according to Heisenberg's interpretation, and to the Copenhagen interpretation, a situation very much like this arises in *all measurements*, so that formula (1) holds for *all measurements* of one variable and a non-commuting second variable, for all elementary particles, for example, photons or electrons.

This was not at first understood to be part of the formalism itself (for example, the Schrödinger equation). It was, at first, shown by Heisenberg in applying the formalism to a small number of physical situations which could be described as *measurements*.

Heisenberg tried to *explain* the limitations which his interpretation imposed on all possible measurements by pointing out that if we measure an elementary particle, we disturb it or interfere with it.

This early interpretation of Heisenberg's implied that the particle *had* a sharp position *and* momentum; but owing to our interfering with it, we could never measure both sharply.

17

This interpretation was changed after Schrödinger suggested that a particle may be represented by a wave packet, or may indeed *be* a wave packet.[23]

The argument of Einstein, Podolsky and Rosen can be regarded as directed against (a) the view that a particle cannot at the same time *have* a sharp position and momentum and also against (b) the view that *every* measurement of position must interfere with the particle's momentum, and *vice versa*. Put very briefly, it proceeds as follows.[24] Suppose one has a composite system described by a Schrödinger equation, containing, say, two particles, A and B, prior to any collision between them. Then they collide; and afterwards one of the particles is measured, which we may call A. We may choose to measure various aspects; for example, we may measure position; or we may measure momentum. If the position of A is measured, this, together with the ψ-function of the composite system, will allow us to find the position of B. If the momentum of A is measured, one can similarly obtain the momentum of B. As Einstein put it: 'Quantum mechanics will then [that is, after measuring A] give us the ψ-function for the partial system B, and it will give us *various ψ-functions* [of B] *that differ, according to the kind of measurement which we have chosen to carry out upon A.*'[25] B may meanwhile have moved as far away as one likes—to Sirius, say. In other words, contrary to the original teaching of the Copenhagen interpretation, one can predict either the position or the momentum of B, without in any way interfering with or disturbing B, on the basis of a measurement (and disturbance) of A alone.

[23]Natural laws prohibit certain kinds of events, and the Heisenberg principle does so too: certain logically possible physical events—say, a beam of light passing through a slit without scatter—would logically *contradict* the theory. There were always people who liked to call such forbidden events 'meaningless'; but this is a mistake: we should certainly distinguish between at least three very different things: logically possible events forbidden by a theory; logically impossible events (or events forbidden by logic); and meaningless pseudo-statements. Preoccupation with the latter was extremely fashionable in the 1920s and early 1930s, mainly due to Wittgenstein's *Tractatus* and its influence on the Vienna circle. I suspect that it was this influence which made John von Neumann attempt to construct a language whose logic excluded as meaningless the formulae forbidden by Heisenberg. I do not see that such a language is (or would be) useful—or that it can solve any physical problems; and I think that the philosophical problems are only confused thereby.

[24]See Einstein's excellent brief statement of the EPR argument in his letter to me of September 11, 1935, printed in *L.Sc.D.*, pp. 457–64.

[25]*Ibid.*, p. 459.

Indeed, B is too far away to be interfered with. Now in EPR it is *assumed* that there is no action at a distance (an assumption that follows from special relativity), and the EPR argument indeed depends upon this assumption. This assumption was later called 'the principle of locality', or 'the principle of local action'. From the fact that we can, as we wish, obtain the position or the momentum of B without even measuring it directly (one measures only A), Einstein concludes that B must have both position and momentum, and that in either of the two choices—whether a measurement of position is chosen or one of momentum—quantum theory allows us only incomplete information about B. And it cannot be our disturbing B that limits the possible information about B—for we do not disturb B. Measuring (or interfering with) A's momentum may destroy A's position; measuring (or interfering with) A's position may destroy A's momentum. But we cannot be doing that also to B—which may be light years away from A, and thus unaffected by measurements of A (unless of course there *is* action at a distance—in fact, instantaneous action, faster than the speed of light, for the measurement of A provides us with information about B for the same moment of time as the moment at which we have measured A).

According to the locality principle, separated and noninteracting objects are independent. Thus B must have an objective reality apart from any 'act of observation', and it must have a sharp position and a sharp momentum at the same time, even though we cannot know both at the same time.

(I may say parenthetically that the EPR paper, important as it is, is in my opinion flawed in certain ways that have distorted the discussion about these matters. For example, too much attention is paid in it to the question of what we should call '*real*'; and too much attention is paid to the question of whether quantum mechanics is complete—a question that is, in my opinion, a red herring in this context in the way in which it is usually discussed, forgetting that behind it there stands what I have called 'the end-of-the-road thesis' in section III above:[26] every physical theory is incomplete in more than one sense, and the incompleteness of quantum mechanics is obvious—since it gives no clue, for example, to the relationships

[26][See Popper's remarks in 'Quantum Mechanics Without "the Observer"', reprinted as the Introduction to this volume; and also Chapter 2, section 9, below. Ed.]

between the absolute values of the physical constants e, c, and h.)

The most important questions raised by the EPR paper are twofold: (1) whether the interpretation of Heisenberg (and similar interpretations) is correct: namely, whether indeterminacy can be explained as due to interference on the part of the observer with the object observed; and (2) whether there is action at a distance or not. Is it possible, by doing something to A—by measuring or observing A (which is close at hand, as it were)—to affect B which may be light years distant?

Now action at a distance is (as mentioned before) excluded by Einstein's special theory of relativity (which, incidentally, is experimentally well corroborated). Special relativity implies the locality principle. If there were action at a distance (even if this action cannot be used for signalling[27]), then special relativity would have to be emended and adjusted accordingly. (In fact, we should have to return to Lorentz's interpretation of the formalism of special relativity, for which see the discussion below.) Neither Einstein nor Bohr (nor Heisenberg) ever suggested that locality should be abandoned. And I believe that, had Bohr been confronted with theoretical results suggesting that quantum theory implied the abandonment of locality, he would have regarded this as an extremely strong argument against quantum theory, and perhaps even as a refutation of it.[28]

This becomes clear from a rather little-known exchange between Einstein and Bohr which was published in *Dialectica* in 1948.[29] In

[27]It is often implied by physicists that only the possibility of *sending signals* whose velocity exceeds that of light would refute special relativity. This, however, is not correct. The moment we say that two distant events are simultaneous in an absolute sense we have abandoned Einstein's relativistic interpretation of the Lorentz-Einstein formalism. For within special relativity, two events on the x axis which are simultaneous relative to the inertial system S_1 are never simultaneous relative to the inertial system S_2 unless S_1 and S_2 are not moving relative to each other along the x axis, even if there should be no interaction (and therefore no signalling) between these two events.

[28]We should remember that Bohr appealed to general relativity in defending the general validity of the Heisenberg relations; see P. A. Schilpp, ed.: *Albert Einstein: Philosopher-Scientist*, pp. 225–28. It cannot be sufficiently stressed that, apart from Einstein in his *Dialectica* article, nobody before Bell suspected that quantum mechanics itself (and not perhaps only the Copenhagen interpretation) clashed with locality, and therefore with special relativity.

[29]Albert Einstein: 'Quantenmechanik und Wirklichkeit', *Dialectica* **2**, 3–4, November 1948, pp. 320–24.

his *Dialectica* article Einstein advances a very modest and simple argument against the Copenhagen interpretation of quantum theory. First he formulates clearly the principle of locality, as it is now called, that is, the principle of the excluded action at a distance, referring to it as the principle of action at vanishing distances, or the *'Prinzip der Nahewirkung'*. He then notes that the principles of quantum mechanics, at least in the Copenhagen interpretation, are incompatible with this principle of locality, and that if quantum mechanics is true as interpreted by Bohr there must be action at a distance.

He goes on to say that 'it seems to me quite clear that those physicists who regard the ways of description of quantum mechanics as in principle final *('definitiv')* will react to these considerations as follows: they will drop the requirement... of the independent existence of physical real things in distant parts of space; and they could rightly claim that quantum mechanics nowhere makes implicit use of any such requirement'.

Einstein immediately admits this claim: he admits that the principle of locality is not an explicit thesis of quantum theory. But he goes on to caution us that, when he considers all the physical phenomena which are known to him, and especially those that are successfully described by quantum mechanics, he nevertheless nowhere finds a physical fact that would make it probable for him that the principle of locality has to be abandoned. He concludes that 'In any case one should, in my opinion, in the course of searching for a unitary basis for the whole of physics, not become dogmatically attached *(dogmatisch festlegen)* to the formulations of contemporary theories.'

If one consults Bohr's reply to Einstein's paper[30] (which, incidentally, although written later, and with Einstein's paper in front of him, was printed by Pauli before Einstein's paper) or Pauli's own editorial introduction to the discussion, one gets the clear impression that neither of them thought that Einstein interpreted quantum theory correctly when he maintained that it implied action at a distance. Pauli dismisses as 'meaningless' any statement about simultaneous values of observables to which no possible vector in

[30][See also Popper's discussion of Bohr's replies to Einstein in *L.Sc.D.*, Appendix *xi. Ed.]

Hilbert space exists, and seems to think that this is a sufficient answer to Einstein. And if one reads Bohr's paper, one finds that he nowhere admits, and upholds against Einstein, that in quantum mechanics there is indeed action at a distance.

VII

As a result of the developments of the past decade, however, all of this seems to have changed. Many contemporary physicists would say that what we might call 'the Einsteinian alternative'—either quantum mechanics or the principle of locality—was indeed correctly drawn (though they do not seem to know that it was drawn by Einstein), which implies that Bohr's argument against Einstein's analysis was incorrect, and even incoherent; *but they think that nevertheless Einstein was wrong in holding on to the principle of locality; and that Bohr's non-realist philosophical position was correct.*

These new developments reach back to David Bohm's restatement of EPR in terms of spin, rather than position and momentum. Bohm's version of EPR (which, from my point of view, differs significantly from the original EPR) supposes that there are two particles whose spins interact. Afterwards the spin of one is measured, and thereby the spin of the other is obtained. Different components of the spin of, say, A may be measured, and we thereby obtain the corresponding components of the spin of B.[31]

I may perhaps mention here some of the differences between the original EPR argument and Bohm's version of it. These differences relate to the distinction of two kinds of quantum-mechanical state

[31]Bell pointed out that in Bohm's version of EPR a crucial experiment is possible with respect to a polarized particle B. That is, when we do something to particle A, something should, according to quantum theory, happen simultaneously to particle B; whereas according to a commonsense realistic and local theory, particle B would not be affected—at least not simultaneously. These experiments are not merely crucial between the Copenhagen interpretation and the quantum theoretical formalism. On the contrary, in these experiments we have ranged on one side the quantum theoretical formalism and the Copenhagen interpretation, and on the other side local theories, and hence special relativity theory. [See John F. Clauser and Abner Shimony: 'Bell's Theorem: Experimental Tests and Implications', *Reports on Progress in Physics*, 1978. See also Popper's discussion of Bohm in *L.Sc.D.*, Appendix *xi, esp. pp. 447 *ff*. Ed.]

preparations.[32] State preparations either may be merely selective, or may in addition impose new propensities—as in the case of selecting a state of a particle by way of a polarizer (for example, a Nicol prism or a tourmaline). The first kind is connected either with sheer retrodictive experiments of measurement or with the sort of state preparation that I have called 'physical selection'.[33] The Bohm (and hence also the Bell) version of EPR, however, is selective in the second sense: these experiments are based on polarization measurements which impose new propensities on the particles in question. Hence there is at least a possibility that an experiment with this type of state preparation may differ significantly from an experiment with the other kind of state preparation. Indeed, it is possible that the Bohm-Bell experiment decides for action at a distance, and therefore against special relativity theory, whereas the original EPR argument does not. Should this be so, then we would have a very new and interesting difference between two types of selection or state preparation.

(It may be mentioned parenthetically that there are some other problems with regard to spin. First, we know so little about spin that it could be that Bohm is mistaken in supposing that his version of EPR, relating to spin, is essentially equivalent with the form

[32]I believe that the idea of state preparation in the sense in which it is now used in quantum mechanics was introduced by me in 1934 in *L.Sc.D.*; see especially pp. 225–26. I there called it 'physical selection', and I distinguished it as a kind of predictive measurement from those retrodictive measurements (like the recording of a particle on a photographic plate) which may strongly interfere with a particle and even destroy it, and of which Heisenberg had suggested that they may be, because of their non-predictive character, meaningless. Henry Margenau referred to state preparations in a paper in 1937.

[33]The two kinds of state preparations discussed here should not be confused with the two kinds of measurement (retrodictive and predictive—the latter are state preparations) mentioned in the foregoing footnote. The predictive measurements or state preparations produce, in quantum mechanics, always a scatter of those 'observables' which are characterized by variables which do not commute with those variables which characterize the prepared state. A measurement (or state preparation) of spin is indeed of the 'mousetrap' character in the sense that I introduce in *L.Sc.D.*, Appendix *xi, p. 453. While true for spin, this is not true for position and momentum. Only spin has this queer property. That is, the mousetrap holds for some state preparations and not for others. See my paper 'Particle Annihilation and the Argument of Einstein, Podolsky, and Rosen', in Wolfgang Yourgrau and Alwyn van der Merwe, eds.: *Perspectives in Quantum Theory*, 1971, esp. pp. 187–189.

23

proposed by Einstein. My reason for saying this is that we have no realistic theory of spin. Spin, as it occurs in atomic theory, differs from what we usually, realistically, mean by spin. Spin is really something very queer and is *in a sense* non-classical. Perhaps with spin one actually does have action at a distance (see below). But not necessarily for every distance: with big distances it may not work. But this is crucial for the original EPR. The original argument was *meant* to make use of great distances. Thus if we find evidence for a short-distance 'action at a distance', this may not really be telling against the original version of EPR, but only against the argument of EPR in Bohm's version. Thus the arguments concerning spin may not decisively help the Copenhagen interpretation after all. These new experiments would save the Copenhagen interpretation only if non-local action would work over distances of any size. Incidentally, the necessity of distinguishing between the formalism of quantum mechanics and the Copenhagen interpretation is even now seldom realized.)

However this may be, once the EPR experiment is restated in terms of spin, it need not merely be a thought experiment; it can actually be tested. The theoretical basis for such an experiment was worked out by J. S. Bell; and a number of such experiments have now been carried out.[34] (My remarks in the following are based on

[34][See J. S. Bell: 'On the Einstein Podolsky Rosen Paradox', *Physics* 1, 3, pp. 195–200, 1964; 'On the Problem of Hidden Variables in Quantum Mechanics', *Reviews of Modern Physics* 38, pp. 447–452, 1966; *Epistemological Letters* 15, Bienne, France, pp. 79–84. For the experiments, see S. J. Freedman and J. F. Clauser: 'Experimental Test of Local Hidden-Variable Theories', *Physical Review Letters* 28, pp. 935–941, 1972; R. A. Holt and F. M. Pipkin: *Quantum Mechanics vs. Hidden Variables: Polarization Correlation Measurement on an Atomic Mercury Cascade*, 1973; John F. Clauser: 'Experimental Investigation of a Polarization Correlation Anomaly', *Physical Review Letters* 36, pp. 1223–1226, 1976; E. S. Fry and R. C. Thompson: 'Experimental Test of Local Hidden-Variable Theories', *Physical Review Letters* 37, pp. 465–468, 1976; G. Faraci, S. Gurkowski, S. Natarrigo, and A. R. Pennisi: 'An Experimental Test of the EPR Paradox', *Lettere al Nuovo Cimento* 9, pp. 607–611, 1974; L. R. Kasday, J. D. Ullman, and C. S. Wu: 'Angular Correlation of Compton-scattered Annihilation Photons and Hidden Variables', *Nuovo Cimento*, 25B, 1975 (ser. 2, no. 2), pp. 633–661; M. Lamehi-Rachti and W. Mittig: 'Quantum Mechanics and Hidden Variables: A Test of Bell's Inequality by the Measurements of the Spin Correlation in Low Energy Photon-proton Scattering', *Physical Review* D, 14, pp. 2543–2555, 1976. See K. R. Popper, A. Garuccio, J.-P. Vigier, 'An Experiment to Interpret E.P.R. Action-at-a-Distance: The Possible Detection of Real De Broglie Waves', *Epistemological Letters*, 30, July 1981, pp. 21–29. Ed.]

the assumption that Bell's *interpretation* of his theorem is physically correct; this is, however, an open question.)

These tests involve what is now called the Bell inequality. Although the results of testing so far are not yet quite conclusive, most of them seem to go against what Bell calls 'local realistic theories', and in favour of quantum theory and perhaps even its Copenhagen interpretation.

I must admit that these test results have surprised me. When I first heard that John F. Clauser and Abner Shimony were intending to test Bell's theorem, I expected that their results would refute quantum theory. But my expectation appears to have been mistaken, since the majority of the tests have gone the other way.

VIII

Nevertheless I have not abandoned my realistic interpretation of physics; and so far I have not even abandoned locality.

On the contrary, I think (in contrast to Abner Shimony) there is not the slightest reason to suppose that realism is affected by these new experiments, even if their outcome should show that locality cannot be upheld. Rather, as I explain in the next section, the consequences, if correct, would go against Einstein's interpretation of the formalism of special relativity theory and in favour of Lorentz's interpretation of it and in favour of Newton's 'absolute space'. (And both Lorentz and Newton were of course realists.)

These new experiments based on Bell's theorem would indeed be the first that could be taken as crucial between Lorentz's theory and special relativity theory—although they were not designed as such, and although no one (as far as I know) has said that they could be used in this way in order to decide between Lorentz and Einstein.

For a long time, and quite independently of the EPR argument, I have said that although the idea that there is no action at a distance is in every respect intuitively more satisfactory—especially for a realist—than is the idea of action at a distance, we should not think that we have any knowledge *a priori* that action at a distance is impossible, or that it is excluded by realism.

This view differs a bit from Einstein's, who has written: 'But there is *one* assumption which, in my opinion, we should retain under all circumstances: the real factual state of the system S_2 is independent

of what is done to the system S_1, where S_1 is a system that is spatially separated from S_2.'[35] I do think we should not give up this assumption easily; but we need not retain it 'under all circumstances'. We should be prepared to entertain the possibility that there is action at a distance after all.

But before discarding so intuitively satisfactory a principle as that of local action, the whole situation must be reviewed much more deeply and comprehensively than has been done so far. Of course, as I have already mentioned, action at a distance is incompatible with special relativity; that is to say, with Einstein's (as opposed to Lorentz's) interpretation of its formalism. And it would, if accepted, demand a considerable modification of special relativity theory. I do not mean to suggest that special relativity theory—that is, Einstein's interpretation—would have to be thrown away. But we should remember that special relativity is to be regarded, from the point of view of general relativity theory, as a first approximation only.

In any case, the main thing that I want to say—and that, as I believe, Einstein wanted to say—is this. The normal way in which things happen in this world is in accordance with local action and thus quite contrary to the apparent outcome of these experiments. If we are to accept action at a distance, we should have to allow for an abnormal as well as a normal way for things to happen in the world. That would be a major blow against common sense. But all our commonsense ideas, including this one, should always be open to criticism.

Moreover, it is not just common sense that conflicts with these experiments and with the rejection of locality. Everything that we know from astronomy and from the technical success of physics also conflicts with them: they all suggest the reality of time and the exclusion of action at a distance. Even more important, the idealistic consequences—especially, the theory that the flow of time is a subjective delusion—that are being drawn from these experiments and from the whole situation in atomic physics seem to me to conflict very significantly with biology and with the theory of evolution.[36]

[35]See P. A. Schilpp, ed.: *Albert Einstein, Philosopher-Scientist*, 1949, p. 85. The translation is slightly revised here.

[36][See Popper's remarks in Peter Medawar and Julian Shelley: *Structure in*

We shall, then, require a great deal more than these recent experiments, about which many questions in any case have to be answered, before we reject locality. And, again, they in no way—despite what is frequently asserted—conflict with realism. Philosophers and physicists have been all too prone during the past sixty years, largely under the indirect influence of Machian positivism, to take up idealist positions. One of the things that this volume of the *Postscript* tries to do is to review many of the *past* arguments for idealism—which many current physicists still simply take for granted—and to show their error.

IX

I wish to propose a simple experiment which may be regarded as an extension of the Einstein-Podolsky-Rosen argument.[37] As originally formulated, the EPR 'thought experiment' is only an argument, not a real experiment. I wish to suggest a crucial experiment to *test* whether knowledge alone is sufficient to create 'uncertainty' and, with it, scatter (as is contended under the Copenhagen interpretation), or whether it is the physical situation that is responsible for the scatter.

We have a source S (positronium, say) from which pairs of particles that have interacted are emitted in opposite directions. We consider pairs of particles that move in opposite directions along the positive and negative x-axis, towards two screens A and B, with slits whose width Δq_y is adjustable (compare Figure 2, p. 28). Beyond the slits on both sides there are batteries of Geiger counters arranged in semi-circles.

We assume that the intensity of the beam of the emitted particles is very low, so that the probability is high that two particles recorded at the same time on the left and on the right are those which have actually interacted before emission.

Those particles that passed through the slits A and B will be

Science and Art, Proceedings of the Third C. H. Boehringer Sohn Symposium, Kronberg, Taunus, Amsterdam, 1980, pp. 154–168, esp. p. 167. See also W. W. Bartley, III: 'Philosophy of Biology *versus* Philosophy of Science', *Fundamenta Scientiae* 3, 1, 1982. Ed.]

[37]The experiment could in principle be made with pairs of photons created by the annihilation of particles.

Figure 2

counted by Geiger counters. These counters are coincidence coun-
ters; that is to say they are so wired that they only count particles
that have passed at the same time through A and B. This should
make it almost certain that only pairs of particles which have
interacted are recorded.

We now first test the Heisenberg scatter for both the beams of
particles going to the right and to the left, by making the two slits A
and B wider or narrower. If the slits are narrower, then counters
should come into play which are higher up and lower down, seen
from the slits. The coming into play of these counters is indicative of
the wider scattering angles which go with a narrower slit, according
to the Heisenberg relations.

Now we make the slit at A very small and the slit at B very wide.

According to the EPR argument, we have measured q_y for both
particles (the one passing through A and the one passing through B)
with the precision Δq_y of the slit at A, since we can now calculate the
y-coordinate of the particle that passes through B with approxi-
mately the same precision, even though its slit is wide open.

We thus obtain fairly precise '*knowledge*' about the q_y position of
this particle—we have 'measured' its y position indirectly. And
since it is, according to the Copenhagen interpretation, our *knowl-
edge* which is described by the theory—and especially by the
Heisenberg relations—we should expect that the momentum p_y of
the beam that passes through B scatters as much as that of the beam
that passes through A, even though the slit at A is much narrower
than the widely opened slit at B.

Now the scatter can, in principle, be tested with the help of the counters. If the Copenhagen interpretation is correct, then such counters on the far side of B that are indicative of a wide scatter (and of a narrow slit) should now count coincidences: counters that did not count any particles before the slit at A was narrowed.

To sum up: if the Copenhagen interpretation is correct, then any increase in the precision of our *mere knowledge* of the position q_y of the particles going to the right should increase their scatter; and this prediction should be testable.

If, as I am inclined to predict, the test decides against the Copenhagen interpretation, then this does not mean that quantum mechanics (say, Schrödinger's formalism) is undermined; although it would mean that Heisenberg's claim is undermined that his formulae are applicable to all kinds of indirect measurements (a claim which the adherents of the Copenhagen interpretation—such as von Neumann—would undoubtedly regard as part of quantum mechanics).

What would be the position if our experiment (against my personal expectation) supported the Copenhagen interpretation— that is, if the particles whose y-position has been indirectly measured at B show an increased scatter?

This *could* be interpreted as indicative of an action at a distance; and if so it would mean that we have to give up Einstein's interpretation of special relativity and return to Lorentz's interpretation and with it to Newton's absolute space and time. We need not, in that case, give up any formula of special relativity theory. For special relativity is an interpretation of a formalism; and the same formalism can be interpreted either by special relativity or by Lorentz's view that we have an absolute space and time but cannot detect it, for reasons that are revealed by the formalism. Whereas special relativity theory, in Einstein's interpretation, says that simultaneity has no absolute sense; that if we do not have any way of detecting absolute space and time—if its detection is indeed excluded by the formalism—then we should not assume that it exists.

Of course in those days, Lorentz's adherence to an ether at rest and Newtonian absolute space and time was a bit shocking, whereas the relativity interpretation of the formalism was simple, elegant, and convincing. The decisive thing about Einstein's theory, from

my point of view, is that it has shown that Newton's theory—which has been more successful than any other theory ever proposed—*can* be replaced by an alternative theory which is of wider scope, and which is so related to Newton's theory that every success of Newtonian theory is also a success for that theory, and which in fact makes slight adjustments to some results of Newtonian theory. So for me, this logical situation is more important than the question which of the two theories is in fact the better approximation to the truth.

Bohr too was, of course, a passionate admirer of special relativity theory. He would have wanted to avoid rejecting it—like almost everybody in those days. Had it been shown that such a rejection would be necessary if we wish to uphold quantum mechanics, it may well have meant, even for Bohr, the rejection of quantum mechanics. For special relativity more or less set the standard to which quantum mechanics had to conform.

It is only now, in the light of the new experiments stemming from Bell's work, that the suggestion of replacing Einstein's interpretation by Lorentz's can be made. If there is action at a distance, then there is something like absolute space. It should be remembered that nothing like a crucial experiment between Lorentz's theory and special relativity theory has ever been conceived before. If we now have theoretical reasons from quantum theory for introducing absolute simultaneity, then we would have to go back to Lorentz's interpretation.[38]

X

This volume of the *Postscript* bears mainly on problems of physics while, at the same time, leading up to what I call a 'Metaphysical Epilogue', which sketches the outline of what seems to me to be a new and promising way of looking at the physical cosmos, and especially at the problem of matter; a way which, I hope, might be able to solve some of the major difficulties of quantum theory. This

[38]There are independent arguments for such a course, especially since the discovery of the microwave cosmic background radiation. See E. K. Conklin: *Nature* 222, 1972, p. 971. [Popper's suggestion that experiments establishing non-locality could be used as crucial between Lorentz and Einstein dates from June 1980. Ed.]

volume, then, was inspired not so much by microphysical theory, important as that is, but by the problems of physical cosmology.[39]

Cosmological speculations play, and have always played, an immensely important role in the development of science. By examining and probing these speculative theories, we may be able to develop them greatly, and perhaps eventually even bring them to the point where they may be tested empirically. (This certainly has been done by the authors of the steady state theory of the expanding universe: the tests led to its abandonment.) Many cosmological speculations—particularly in their initial formulations—cannot be tested or empirically falsified; so I prefer to call them 'metaphysical' rather than 'scientific'.

In this volume I introduced the term 'metaphysical research programme' in order to refer to the twofold character of these important cosmological theories: their programmatic character, often shaping and determining the course of scientific research and development; and their (at any rate at first) untestable, and thus metaphysical, character.

In the 'Metaphysical Epilogue' of the present volume, I said little about research programmes in general. Rather, I proposed a real research programme for application to and interpretation of real problems in physics and biology. In the course of introducing my own metaphysical research programme, I had to examine the history of science, and to reconstruct the 'logic of the situation'; that is, to articulate the successive problem situations in physics (and later in biology) in terms of which my own metaphysical research programme, and its many precursors and competitors, had to be seen in order to be understood, criticized, and evaluated.

Such metaphysical research programmes may be found everywhere. They determine, or arise from, what is regarded as a satisfactory explanation at any moment of the evolution of a scientific problem. It seems that what Thomas S. Kuhn was to call 'paradigms'[40] is somewhat similar, although of course my point of

[39]This interest has, in the years since I first wrote this part of the *Postscript*, led me beyond physics, especially to biology, to the human mind, and to the products of the human mind (which I have called World 3). See *Objective Knowledge*, 1972; *The Self and Its Brain*, 1977; and also the Afterword and Addenda to Volume II of the *Postscript*.

[40]Thomas S. Kuhn: *The Structure of Scientific Revolutions*, 1962.

view is radically different from his: I look upon these programmes in terms of a situation that can be rationally reconstructed, and see scientific revolutions as due, essentially, to rational criticism—that kind of activity which I regard as most important in the history of science. Admittedly, the refutation of a scientific theory may often have an impact on the scientist (as for example in the case of Rutherford) similar to a conversion; but it would be a rational conversion.[41]

The name 'metaphysical research programmes' was used by me for certain research programmes for science; that is, for those which are not yet testable. More than ten years later, some of my collaborators changed the name into 'scientific research programmes'.[42] These research programmes certainly pertain to science, and play a crucially important role in the development of science. But they are usually not yet of the character of testable scientific theories. They may *become* scientific theories, but then there is no point any longer in calling them research programmes. They are in any case much harder to criticize than are theories—and much easier to retain uncritically.

Actually, there is no need to praise the merits of research programmes (let alone to develop a 'methodology' for them).

[41]See my 'The Rationality of Scientific Revolutions', The Herbert Spencer Lecture for 1973, in Rom Harré, ed.: *Problems of Scientific Revolution: Progress and Obstacles to Progress in the Sciences*, 1975.

[42][See Imre Lakatos: *The Methodology of Scientific Research Programmes*, 1978. Lakatos knew Popper's *Postscript* and possessed a copy of the original galley proofs of 1957. In one place (*Criticism and the Growth of Knowledge*, 1970, ed. Lakatos and Musgrave, p. 183 *f.*), he refers to the *Postscript* and quotes at length from it. (In this discussion he distorts Popper's view, accusing Popper of neglecting 'methodological irrefutability' for the sake of 'syntactical irrefutability'. Popper distinguishes these in *L.Sc.D.*, Chapter IV, esp. section 20.) In another publication—'Changes in the Problem of Inductive Logic', in I. Lakatos, ed.: *The Problem of Inductive Logic*, 1968, p. 316—Lakatos introduces the idea of a research programme without crediting Popper at all. Perhaps as a consequence, Popper's alleged views are now often presented as the 'orthodox' position, in contrast and opposition to Lakatos's 'methodology of scientific research programmes'. See for example Peter Urbach: 'The Objective Promise of a Research Programme', in Gerard Radnitzky and Gunnar Andersson, eds.: *Progress and Rationality in Science*, 1978, pp. 99 *ff.* See W. W. Bartley, III's discussion of the name change in his 'On Imre Lakatos', in R. S. Cohen, P. K. Feyerabend, and M. W. Wartofsky, eds.: *Essays in Memory of Imre Lakatos*, 1976, pp. 37–8. See also his discussion in 'The Philosophy of Karl Popper: Part III: Rationality, Criticism, and Logic', *Philosophia*, Israel, February 1982. Ed.]

Scientists are fitted by their whole education into research programmes (so far one might say that Kuhn is right). There is far more need to worry about getting stuck in or 'programmed by' a research programme. There is as a rule far more need to encourage people to criticize, and to break out of, research programmes than to encourage them to adopt a research programme and to adhere to it—unless, of course, one has a new research programme to sell.

(There is a simple explanation for this in terms of situational logic. Unless he is quite unusually gifted and also unusually well taught, the young scientist may feel so overwhelmed by the richness of the field he tries to enter, and overtaxed by the tasks he sees, that he despairs of his ability to make a contribution to it. His task is made much easier by a research programme. This situational explanation may be regarded as a very partial support of Kuhn's idea of the 'normal' scientist.)

Usually, we become conscious of our research programme only when it dawns on us that it may be based on a *false* metaphysics. To *realize* that we are working within a certain metaphysical research programme is, essentially, to realize that alternatives are possible; and this very realization means that we give up our metaphysical research programme as *the* heuristics, and consider alternatives as possibly more fruitful.

In those rare cases in which the metaphysical research programme does become conscious (or is even deliberately invented, as in the case of the propensity cosmology) it should be recommended only with a warning that we ought to be highly critical of it and look out for possible alternatives.

This is an issue of the rationality of man and, more especially, of the rationality of science. We should always beware of becoming 'normal scientists': scientists who work blindly, uncritically, within the unconscious presuppositions of a research programme. A 'normal scientist' is not attempting to be as rational as he can be, for he is not trying to be as critical as he can be.

The methodologists of scientific research programmes seem to me to show insufficient understanding of the fundamental role played by criticism in the growth of knowledge.[43] As I see it, criticism is the

[43]See I. Lakatos and Alan Musgrave, eds.: *Criticism and the Growth of Knowledge*, 1970.

prime *duty* of the scientist and of anyone who wants to advance knowledge. Seeing new problems and having new ideas, on the other hand, are *not* one's duty: originality is, rather, a gift of the gods.

Indeed, the fascination with research programmes manifested in this latest fad is deeply connected with the old craving for originality, and with the craving for a heuristic, an algorithm, a 'scientific method' that will lead to new and better ideas.[44] As I have hinted elsewhere, no such heuristic, no positive methodology, seems to exist. As to the old question of method: why some people have good ideas and other people not, the proper answer seems to me that some people have many ideas—a few of which may perhaps be good—and are highly critical of their own ideas; whereas others have only a few ideas, and are therefore not very critical of them. One can perhaps increase the number of ideas which one has, and one can certainly increase one's critical attitude towards all one's ideas. But in any case we can never be sure that we are going in the right direction. Even if there did appear to be such a heuristic—even if there were a treadable way towards new ideas or new problems that seemed to be 'good' or 'progressive'—it would remain one of our main tasks to be critical of that very way. We may well be on the wrong way even when—and especially when—we think we are progressing. Even Einstein seems to have been on the wrong way (as I shall explain later in this volume; see section 26) with his own research programme of unifying electromagnetic and gravitational field theories.

[44]See *Realism and the Aim of Science,* Vol. I of the *Postscript,* Introduction, from which the reader will see that I have for many years been arguing against the possibility of such a method or heuristic.

INTRODUCTION: QUANTUM MECHANICS WITHOUT 'THE OBSERVER'

> If somebody asked 'Who is the greatest modern physicist after Einstein?' the answer would be: Einstein again . . . [For] had somebody else discovered relativity, his other discoveries would still make him the second greatest physicist of his time.
>
> CORNELIUS LANCZOS

1. *Quantum Theory and the Role of 'the Observer'.*

In this introduction, I attempt to exorcize the ghost called 'consciousness' or 'the observer' from quantum mechanics, and to show that quantum mechanics is as 'objective' a theory as, say, classical statistical mechanics.[1] In the body of this volume, I shall attempt to substantiate my argument in somewhat greater detail, and to state my own understanding of these issues that have plagued quantum theory over the past fifty years, and my own alternative approach.

My thesis in this introduction is that the observer, or better, the experimentalist, plays in quantum theory exactly the same role as in classical physics. *His task is to test the theory.*

The opposite view, usually called the *Copenhagen interpretation of quantum mechanics,* is almost universally accepted. In brief, it says that *'objective reality has evaporated',* and that *quantum mechanics does not represent particles, but rather our knowledge, our observations, or our consciousness, of particles.*[2]

[1][This introductory chapter is a revised and expanded version of a paper published in *Quantum Theory and Reality,* ed. Mario Bunge, 1967, pp. 7–44. It was written in 1966, approximately ten years later than the body of this volume. The motto is from C. Lanczos: *Albert Einstein and the Cosmic World Order,* 1965, p. 4. Ed.]

[2]*Cp.* Werner Heisenberg, 'The Representation of Nature in Contemporary Physics', *Daedalus* 87, 1958, pp. 95–108; see p. 100.

If a mere philosopher like myself opposes a ruling dogma such as this, he must expect not only retaliation, but even derision and contempt. He may well be browbeaten (though I am happy to remember how kindly and patiently I was treated by Niels Bohr) with the assertion that *all competent physicists know that the Copenhagen interpretation is correct* (since it has been 'proved by experiment').

It seems therefore necessary to point out that this assertion is historically mistaken, by referring to physicists who like Einstein, Planck, von Laue, or Schrödinger, are as competent as any, and who (unlike Einstein, Planck, von Laue, and Schrödinger) were even at one time fully convinced adherents to the Copenhagen interpretation, but who do not now 'regard the new interpretation as conclusive or convincing', as Heisenberg puts it.[3]

There is, first, Louis de Broglie, a one-time adherent to the Copenhagen interpretation; and his former pupil, Jean-Pierre Vigier.

There is, next, Alfred Landé, also one of the great founders of quantum theory in the years 1921 to 1924 who later (1937 and 1951) wrote two textbooks on quantum mechanics entirely in the Copenhagen spirit, but who still later[4] became one of the leading opponents of the Copenhagen interpretation.

There is David Bohm who published in 1951 a textbook, *Quantum Theory*,[5] which was not only orthodox in the Copenhagen sense but one of the clearest and fullest, most penetrating and critical presentations of the Copenhagen point of view ever published. Shortly afterwards, under the influence of Einstein, he tried new ways, and in 1952 arrived[6] at a tentative theory (later revised[7]) whose logical consistency proved the falsity of the constantly repeated dogma (due to von Neumann[8]) that the quantum theory is

[3] W. Heisenberg, 'The Development of the Interpretation of the Quantum Theory', in W. Pauli (ed.), *Niels Bohr and the Development of Physics*, 1955, p. 16.

[4] A. Landé, *Foundations of Quantum Theory*, 1955; *From Dualism to Unity in Quantum Mechanics*, 1960; *New Foundations of Quantum Mechanics*, 1965.

[5] D. Bohm, *Quantum Theory*, 1951.

[6] D. Bohm, 'A Suggested Interpretation of the Quantum Theory in Terms of "Hidden" Variables', *Physical Review* **48**, 1952, pp. 166–79, 180–93.

[7] D. Bohm and J. Bub, 'A Proposed Solution of the Measurement Problem in Quantum Mechanics by a Hidden Variable Theory', *Reviews of Modern Physics* **38**, 1966, pp. 453–69.

[8] J. von Neumann, *Mathematical Foundations of Quantum Mechanics*, 1949, 1955 (German edition 1932).

'*complete*' in the sense that it must prove incompatible with any more detailed theory.

There is Mario Bunge who in 1955 published a paper, 'Strife about Complementarity'.[9]

There is the German physicist, Fritz Bopp, who explicitly subscribes to the Copenhagen interpretation, in an epistemological paragraph of a most interesting paper. In it he writes, for example: 'Naturally our considerations do not mean any alteration of the mathematical concept of complementarity.' Yet he develops there (and in other publications) a theory with which Einstein would hardly have had any quarrel since, on lines not dissimilar to Einstein's, Bopp interprets the quantum-theoretical formalism as an extension of classical statistical mechanics; that is, as a theory of ensembles.[10]

I have given this brief and of course quite incomplete list of dissenters merely to combat the historical myth that only philosophers (and totally incompetent or senile physicists) can doubt the Copenhagen interpretation. But before proceeding to criticize this interpretation in some detail, I should like to discuss two points.

(a) In a very important sense, which to my knowledge has been usually overlooked, the Copenhagen interpretation ceased to exist long ago.

(b) Most physicists who quite honestly believe in it do not pay any attention to it in actual practice.

As to point (a), we must not forget that 'the new quantum theory' or 'quantum mechanics' was, to start with, and until at least 1935, simply another name for '*the new electromagnetic theory of matter*'.

In order to realize fully how the theory of the atom, and therefore the theory of matter, were identified with the theory of the electromagnetic field, we may for example turn to Einstein, who said in 1920:[11] '. . . according to our present conceptions the

[9] M. Bunge, 'Strife About Complementarity', *British Journal for the Philosophy of Science* 6, 1955, pp. 1–12, 141–54.

[10] F. Bopp, 'Statistische Mechanik bei Störung des Zustands eines physikalischen Systems durch die Beobachtung', in F. Bopp (ed.), *Werner Heisenberg und die Physik unserer Zeit*, 1961, pp. 128–49; see pp. 147 f. See also A. Einstein, 'Remarks Concerning the Essays Brought Together in this Co-operative Volume', in P. A. Schilpp (ed.), *Albert Einstein: Philosopher-Scientist*, 1949, pp. 665–68; cp. pp. 671 f.

[11] A. Einstein, *Sidelights on Relativity*, 1922; cp. p. 22 (the italics are mine).

elementary particles are . . . nothing but condensations of the electromagnetic field . . . Our . . . view of the universe presents two realities . . . , namely, gravitational ether and electromagnetic field, or—as they might also be called—space and matter.'

Quantum mechanics was regarded by its adherents as *the final form of this electromagnetic theory of matter.* That is to say, the formalism was regarded, first of all, as *the theory of electrons and protons* and thereby as *the theory of the constitution of atoms: of the periodic system of elements and their physical properties: and of the chemical bond, and thus of the physical and chemical properties of matter.*

A very impressive statement of the view held by almost all physicists at least up to the discovery of the positron in 1932 is due to Robert A. Millikan:[12]

'Indeed, nothing more beautifully simplifying has ever happened in the history of science than the whole series of discoveries culminating about 1914 which finally brought practically universal acceptance to the theory that the material world contains but two fundamental entities, namely, positive and negative electrons, exactly alike in charge, but differing widely in mass, the positive electron—now usually called a *proton*—being 1850 times heavier than the negative, now usually called simply the *electron.'*

In fact until at least 1935 some of the greatest physicists[13] believed that, with the advent of quantum mechanics, *the electromagnetic theory had entered into its final state,* and that the results of quantum mechanics strongly confirmed that *all matter consisted of electrons and protons.* (Neutrons and neutrinos had also been admitted, somewhat grudgingly, but it was thought that neutrons were protons + electrons; and that neutrinos might not be much more than a mathematical fiction; while positrons were regarded as 'holes' in the sea of electrons.)

This theory that matter consists of protons and electrons died long ago. Its ailment (though it first remained hidden) started with the discovery of the neutron and also of the positron (which the Copenhagen authorities refused to believe in at first): and it received

[12]R. A. Millikan, *Time, Matter and Values,* 1932, p. 46 (the italics are mine); *cp.* also his *Electrons (+ and −), Protons, Photons, Neutrons and Cosmic Rays,* 1935, p. 377.

[13]*Cp.* A. Eddington, *Relativity Theory of Protons and Electrons,* 1936.

its final blow with the discovery of the sharply distinct *levels of interaction*, of which the electromagnetic forces constitute just one among at least four:

1. Nuclear forces (strong interactions).
2. Electromagnetic forces.
3. Weak decay interactions.
4. Gravitational forces.

Moreover, the hope of solving within quantum mechanics such classical problems of the electromagnetic theory as the explanation of the electronic charge has been practically abandoned.

In the light of this situation, we may now look back upon the titanic struggle between Einstein and Bohr. The problem posed by Einstein was whether quantum mechanics was '*complete*'. Einstein said no. Bohr said yes.[14]

I have no doubt that Einstein was right. But even today we can read that it was Bohr who won that famous battle. This view persists largely because Einstein's attack upon Bohr's assertion of the *completeness* of quantum mechanics was interpreted by the Copenhagen school as an attack upon *quantum mechanics itself and its 'soundness' or consistency*. But this entails that we accept (i) the identification of the Copenhagen interpretation with the quantum theory, and (ii) Bohr's shift of the problem from *completeness* to *soundness* (=*freedom from contradiction*). Yet as Einstein had offered his own (statistical) interpretation of quantum theory, he clearly accepted its consistency.

As to point (b), that is, as to my assertion that most physicists who honestly believe in the Copenhagen interpretation do not pay any attention to it in actual practice, an excellent example is Fritz Bopp,[15] since he believes (as do Einstein, Podolsky, and Rosen) that particles possess both sharp positions and momenta at the same time, while the Copenhagen school believes this to be false, or 'meaningless', or 'unphysical'. To quote a formulation of Landé's in

[14]*Cp.* A. Einstein, B. Podolsky, and N. Rosen, 'Can Quantum-Mechanical Description of Physical Reality be Considered Complete?', *Physical Review* **47**, 1935, p. 777–80, and N. Bohr, 'Can Quantum-Mechanical Description of Physical Reality be Considered Complete?', *Physical Review* **48**, 1935, pp. 696–705.

[15]F. Bopp, *op. cit.* (see footnote 10 above).

1951 (before he turned against the Copenhagen interpretation):[16]
'The classical idea of particles breaks down under the impact of the
uncertainty relations. It is unphysical to accept the idea that there
are particles possessing definite positions and momenta at any given
time, and then to concede that these data can never be confirmed
experimentally, as though by a malicious whim of nature.' But what
I have mainly in mind in connection with my point (b) is this.
Admittedly, the formalism of quantum mechanics is still applied by
physicists to the old problems, and its methods are, with many
modifications, partly used in connection with the many new
problems of nuclear theory and elementary particle theory. This is
certainly a great credit to its power. Yet at the same time, most
experimentalists, though much concerned with the limits of preci-
sion of their results, do not seem to be more worried about the role
of the observer or about interfering with their results than they are in
connection with sensitive classical experiments; and most *theorists*
are quite clear that a new and much more general theory is needed:
they all seem to be in search of *a really revolutionary new theory.*

In spite of all this, it still seems necessary *to discuss* the Copenha-
gen interpretation; that is, more precisely, the claim that, in atomic
theory, we have to regard *'the observer'* or *'the subject'* as particu-
larly important, because atomic theory takes its peculiar character
largely from *the interference of the subject or the observer (and his
'measuring agencies') with the physical object under investigation.*
To quote a typical statement of Bohr's:[17] 'Indeed, the *finite
interaction between object and measuring agencies* . . . entails the
necessity of a final renunciation of the classical ideal . . . and a
radical revision of our attitude towards the problem of physical
reality.'

Similarly Heisenberg:[18] '. . . the traditional requirement of sci-
ence . . . permits a division of the world into subject and object
(observer and observed) . . . This assumption is not permissible in

[16]A. Landé, *Quantum Mechanics,* 1951, p. 42. Landé continues by quoting N.
Bohr, 'The Quantum Postulate and the Recent Development of Atomic Theory',
Nature 121, 1928, pp. 580–90.

[17]N. Bohr, 'Discussion with Einstein on Epistemological Problems in Atomic
Physics', in P. A. Schilpp (ed.), *Albert Einstein: Philosopher-Scientist,* 1949; *cp.* pp.
232 *f.*

[18]*Cp.* W. Heisenberg, *The Physical Principles of the Quantum Theory,* 1930,
pp. 2 *f.*

atomic physics; the interaction between observer and object causes uncontrollable large changes in the system [that is] being observed, because of the discontinuous changes characteristic of the atomic processes.' Accordingly, Heisenberg suggests[19] that 'it is now profitable to review the fundamental discussion, so important for epistemology, of the difficulty of separating the subjective and the objective aspects of the world'.

As opposed to all this I suggest that, in practice, physicists measure and experiment today fundamentally in the same way as they did before 1925. If there is an important difference, then it is that the degree of indirectness of measurements has increased as well as the degree of 'objectivity'. Where thirty or forty years ago physicists used to look through a microscope to take a 'reading', such as a count of scintillations, there are now photographic films, or automatic counters, which do the 'reading'. And although photographic films and counter readings have to be interpreted (in the light of theory, as must all experiments or observations) they are in no way physically 'interfered with' or 'influenced' by our theoretical interpretations. Admittedly, many experimental tests now have a largely statistical character, but this makes them no less 'objective': their statistical character (often processed automatically by counters and computers) has nothing to do with the alleged intrusion of the observer, or of the subject, or of consciousness, into physics. By contrast, the preparation or setting up of an experiment always has had, and continues to have, a great deal to do with our changing knowledge: *it depends on theory*.

Our *theories* which guide us in setting up our experiments and in the interpretation of their results have of course always been our inventions: they are inventions or products of our 'consciousness'. But that has nothing to do with the scientific status of our theories which depends on factors such as their simplicity, symmetry, and explanatory power, and on the way they have stood up to critical discussion and to crucial experimental tests; and on their truth (correspondence to reality), or nearness to truth.[20]

[19]Cp. W. Heisenberg, *op. cit.*, p. 65; see also J. von Neumann, *Mathematical Foundations of Quantum Mechanics*, 1949, 1955, pp. 418–21.

[20]Cp. my *Conjectures and Refutations*, 1963, 1972, Chapter 10, and my *Objective Knowledge*, 1972, Chapter 2.

2. *Theories versus Concepts.*

Perhaps this is the best place to insert a few logical remarks on the *distinction between theories and concepts*; remarks which, although what follows does not depend on them, may yet help to remove some obstacles that block the way to a critical understanding of the situation in quantum theory.

What we are seeking, in science, are *true theories*—true statements, true descriptions of certain structural properties of the world we live in. These theories or systems of statements may have their instrumental use: yet what we are seeking in science is not so much usefulness as *truth: approximation to truth; explanatory power, and the power of solving problems;* and thus, *understanding.*

Thus theories are described wrongly if they are described as being *nothing but* instruments (for example, instruments of prediction), though they are as a rule, among other things, also useful instruments. Infinitely more important for the scientist than the question of the usefulness of theories is that of their *objective truth,* or their nearness to the truth, and the kind of *understanding* of the world, and of its problems, which they may open up for us. The view that theories are *nothing but* instruments, or calculating devices,[21] has become fashionable among quantum theorists, owing to the Copenhagen doctrine that quantum theory is *intrinsically ununderstandable* because we can understand only *classical 'pictures',* such as 'particle pictures' or 'wave pictures'. I think this is a mistaken and even a vicious doctrine.

Theories are also described quite wrongly as 'conceptual systems' or 'conceptual frameworks'. It is true that we cannot construct theories without using words or, if the term is preferred, 'concepts'. But it is most important to distinguish between statements and words, and between theories and concepts. And it is important to realize that it is a mistake to think that a theory T_1 is *bound* to use a certain conceptual system C_1: *one* theory T_1 may be formulated in many ways, and may use many different conceptual systems, say C_1 and C_2. Or to put it another way: two theories, T_1 and T_2, should be regarded as one if they are logically equivalent, even though they may use two totally different 'conceptual systems' (C_1 and C_2) or are conceived in totally different 'conceptual frameworks'.

[21]Cp. my *Conjectures and Refutations*, Chapter 3.

I do not happen to believe that Schrödinger[22] and Eckart[23] have validly established the full logical equivalence of wave mechanics and matrix mechanics: there seem to be some loopholes in these equivalence proofs. On this point I agree with Norwood Russell Hanson[24] (and E. L. Hill[25]), although my views on the logic of the equivalence or identity of theories differ somewhat from Hanson's. Yet I do not think that such a proof is impossible, *in spite of the great difference between the conceptual frameworks of the two theories.* (What would be needed for a valid proof is something approaching an axiomatization of both theories, and a proof that to every theorem $t_{1,n}$ of T_1 corresponds a theorem $t_{2,n}$ of T_2 (and *vice versa*) such that, with the help of some system of definitions of the concepts of T_1 and of T_2 we can show that $t_{1,n}$ and $t_{2,n}$ are logically equivalent. It would not be necessary for either T_1 or T_2 itself to contain the means needed for formulating these definitions; for these means may be supplied by some *extensions* of the theories. Incidentally, the fact that definitions may be needed for such an equivalence proof does not mean that they are needed within a physical theory.)

Now since theories *can* be equivalent even though their 'underlying' conceptual frameworks are utterly different (there are many other examples showing that this may be possible), it is clearly a mistake to identify a theory with its 'underlying' conceptual framework or even to believe that these two *must* be very closely related. The conceptual framework of a theory *may* be replaced by a very different one without changing the theory essentially; and *vice versa*: incompatible theories *may* be expressed within the same conceptual framework. (For example, if we replace Newton's inverse square law by an inverse law with the power 2.0001, then we have a different theory within the same framework; and the

[22]E. Schrödinger, 'Über das Verhältnis der Heisenberg-Born-Jordanschen Quantenmechanik zu der Meinen', *Annalen der Physik* 79, 1926, pp. 734–56.

[23]C. Eckart, 'Operator Calculus and the Solution of the Equations of Quantum Dynamics', *Physical Review* 28, 1926, pp. 711–26.

[24]N. R. Hanson, 'Are Wave Mechanics and Matrix Mechanics Equivalent Theories?', in H. Feigl and G. Maxwell (eds.), *Current Issues in the Philosophy of Science*, 1961, pp. 401–25.

[25]E. L. Hill, 'Comments on Hanson's "Are Wave Mechanics and Matrix Mechanics Equivalent Theories?" ', in H. Feigl and G. Maxwell (eds.), *op cit.*, pp. 425–8.

difference will increase if the difference between the two parameters becomes greater. We might even introduce into Newton's theory a finite velocity for gravitational interactions and still say that we are operating within the same conceptual framework. If the velocity is very great, the two theories may be experimentally indistinguishable; if it is small, the theories may differ widely in their empirical implications, though still remaining within the same conceptual framework.)

Although concepts may have great suggestive power, and may thus influence the further development of the theory, it is *not the conceptual system but the theory* that is of real importance for the pure scientist. And the theory is not merely an 'instrument' for him, it is more: he is interested in its truth, or in its approximation to the truth.[26] The conceptual system, on the other hand, is exchangeable and is one among several possible instruments that may be used for formulating the theory. It merely provides a language for the theory; perhaps a better and simpler language than another, perhaps not. In any case, it remains (like every language) to some extent vague and ambiguous. It cannot be made 'precise': the meaning of concepts cannot, essentially, be laid down by any definition, whether formal, operational, or ostensive. Any attempt to make the meaning of the conceptual system 'precise' by way of definitions must lead to an infinite regress, and to merely *apparent* precision, which is the worst form of imprecision because it is the most deceptive form.[27] (This holds even for pure mathematics.)

Thus we are ultimately interested in theories and in their truth, rather than in concepts and their meaning.

This point, however, is rarely seen. Heinrich Hertz said (and Wittgenstein repeated it) that in science we make for ourselves *'pictures'* (*'Bilder'*) of the facts, or of reality; and he said that we choose our 'pictures' in such a way that 'the logically necessary consequences' (*'die denknotwendigen Folgen'*) of the 'pictures' agree with 'the necessary natural consequences' (*'die naturnotwendigen Folgen'*) of the real objects or facts. Here it is left open whether

[26]*Cp.* my *Conjectures and Refutations*, Chapter 10; and the new Appendix II, first published in the 5th, revised, edition of *Objective Knowledge*, 1979.

[27][See *The Open Society and Its Enemies*, 1945, Chapter 11, section 2; and *Unended Quest*, 1974, 1976, section 7. Ed.]

the 'pictures' are *theories* or *concepts*. Mach, in discussing Hertz,[28] suggested that we should interpret Hertz's 'pictures' as 'concepts'. Bohr's view seems to be similar when he speaks (as he so often does) of the *'particle picture'* and the *'wave picture'*; in fact, his way of speaking indicates strongly the (at least indirect) influence of Hertz and Mach.

But 'pictures' are unimportant. They are especially unimportant if they are more or less synonymous with 'concepts', and almost as unimportant when they are meant to characterize theories. *A theory is not a picture.* It need not be 'understood' by way of 'visual images': *we understand a theory if we understand the problem which it is designed to solve, and the way in which it solves it better, or worse, than its competitors.* Some people may combine this kind of understanding with visual images, others may not. But the most vivid visualization does not amount to an understanding of a theory unless these other conditions are realized: an understanding of the problem situation, and of the arguments for and against the competing theories.

These considerations are important because of endless talk about the 'particle picture' and the 'wave picture' and their alleged 'duality' or 'complementarity', and about the alleged necessity, asserted by Bohr, of using 'classical pictures' because of the (admitted but irrelevant) difficulty, or perhaps impossibility, of 'visualizing' and thus 'understanding' atomic objects. But *this* kind of 'understanding' is of little value; and the denial that we can understand quantum theory has had the most appalling repercussions, both on the teaching and on the real understanding of the theory.

In fact, all this talk about pictures has not the slightest bearing on either physics, or physical theories, or the understanding of physical theories. And the fashionable thesis that the attempt to 'understand' modern physical theories is futile since they are essentially 'ununderstandable' (though useful instruments for calculation) amounts to the somewhat absurd assertion that we cannot know what problems they are intended to solve, or why they solve them better, or worse, than their competitors.

If *concepts* are comparatively unimportant, *definitions* must also

[28]E. Mach, *The Science of Mechanics*, fifth edition, 1941 (first German edition 1883); *cp.* p. 318.

be unimportant. Thus although I am pleading here for *realism* in physics, I do not intend to define 'realism' or 'reality'. In pleading here for realism in physics I wish, in the main, to argue that nothing has changed since Galileo or Newton or Faraday concerning the status or the role of the 'observer' or of our 'consciousness' or of our 'information' in physics. I am at the same time quite ready to point out that even in Newton's physics, 'space' was somewhat less real than 'matter' (because although it acted upon matter it could not be acted upon); and that in Einstein's special theory of relativity an inertial frame was less real than a spatio-temporal coincidence of two events, or the spatio-temporal distance between them. In a similar way, the number of degrees of freedom of a physical system is a more abstract idea, and perhaps less real, than the atoms or molecules constituting the system; but still, I should be opposed to saying that the degrees of freedom of a system are not real, that they are *nothing but* a conceptual device, and not a real *physical property of the system*. However, I do not intend to argue about words, including the word 'real', though by and large I regard as excellent Landé's suggestion that we call physically real what is 'kickable' (and able to kick back if kicked). Yet there are, I am inclined to think, degrees of kickability: we can't kick quasars, David Bohm reminds me.[29]

3. *Thirteen Theses.*

I have been in doubt whether I should not perhaps first analyse and criticize the central tenets of the Copenhagen interpretation, and later show that a perfectly realistic interpretation of quantum theory is possible. I have decided to proceed differently. I am going to expound, in the form of thirteen theses and a summary, my own realistic interpretation, for what it is worth; and I shall criticize the Copenhagen interpretation as I go along. I am sure I shall shock many physicists who, after having reached my fourth or at the most my sixth thesis, will stop reading this rubbish: it is to help them not to waste their time that I have decided to proceed as I do.

1. My first thesis concerns the most important thing for under-

[29][See *The Self and Its Brain*, 1977, Chapter P1, section 4. Ed.]

standing quantum theory: the kind of *problems* which the theory is supposed to solve. These, I assert, are essentially *statistical problems*. (a) It was so with Planck's problem in 1897–1900 which led to his radiation formula. (b) It was so with Einstein's photon hypothesis and his derivation of Planck's formula. (c) It was so (at least in part) with Bohr's problem of 1913 which led to his theory of spectral emissions: the explanation of the Rydberg-Ritz combination principle was, clearly, a statistical problem (especially after Einstein's photon hypothesis had been proposed). Admittedly, there was a second problem, thought by Bohr to be *the* fundamental one: the problem of *atomic stability*, or of the stationary ground state of non-radiating electrons in the atom. Bohr 'solved' this problem—by a postulate (of 'quantum states' or 'preferred orbits'). So far as there is any *explanatory* solution to this problem, it is due to de Broglie and to wave mechanics; which *in the light of Born's interpretation* means that it is due to the substitution of a statistical problem for a mechanical problem. (See below.) (d) It was so with the set of problems which were solved first by Bohr's most fruitful 'principle of correspondence': these were, in the main, problems of the *intensities* of the emitted spectral lines. However, Bohr's correspondence arguments were largely qualitative or, at best, approximations. The central problem which led to the new quantum mechanics was to improve on this by obtaining exact statistical results.

This, however, is not at all the way in which Bohr and his school looked at the problem. They did *not* look for a generalization of classical *statistical* mechanics, but rather for a *'generalization of classical* [particle] *mechanics* suited to allow for the existence of the quantum of action', as Bohr put it as late as 1948;[30] a generalization of particle mechanics which would offer 'a frame sufficiently wide to account for . . . the characteristic features of *atomic stability which gave the first impetus to the development of quantum mechanics . . .'*.

Most formulations of the *problem* of quantum mechanics which I have been able to find are similar, except perhaps those 'inductivistic' ones that start from the experiments and look upon theory as 'the attempt to classify and synthesize the results . . . of scientific

[30]N. Bohr, 'On the Notions of Causality and Complementarity', *Dialectica* 2, 1948, pp. 312–19; *cp.* p. 316 (the italics are mine).

experiment',[31] as if the scientific experiments referred to were not, in the main, only the results of theoretical problems, and significant only because of their conflict with, or support of, some theory. (A similar inductivist attitude appears to be Dirac's starting point, when he discusses 'The Need for a Quantum Theory'.[32])

I should admit, however, that Bohr's (in my opinion mistaken) programme of reforming particle mechanics so as to solve the problem of atomic stability appeared to have some prospect of being successfully carried out between 1924 and 1926. I refer, of course, to Louis de Broglie's doctoral thesis of 1923–24 in which he applied to electrons the Einsteinian idea that photons were somehow 'associated' with waves, and showed that Bohr's quantized 'preferred orbits' (and with them, stability) could be explained by wave interference. This was without doubt one of the boldest, deepest, and most far-reaching ideas in this whole development.

De Broglie's idea was, quite consciously, an inversion of Einstein's idea of associating light quanta or photons with light waves. In Einstein's theory, which thus was the model of de Broglie's, light is emitted and absorbed in the form of 'particles' or 'light quanta' or 'photons'; and thus in the form of things which have a pretty sharp spatio-temporal location, at least while they *interact with matter* by being emitted or absorbed. Light is, however, *propagated* like waves. The square of the amplitude of these waves determines, according to Einstein, the density (that is, the statistical probability) of the photons; and the amplitude of the waves at the place where an atom (in an appropriate state) or a free electron is located determines the probability of the absorption of a photon.

However it was more than two years, during which de Broglie's theory of electrons grew into Schrödinger's 'wave mechanics', before Max Born applied to this new wave mechanics the statistical interpretation of the relationship between photons and light waves which we owe to Einstein. Max Born himself says about his statistical interpretation of wave mechanics:[33] 'The solution . . .

[31]Cp. W. Heisenberg, *The Physical Principles of the Quantum Theory*, 1936, p. 1; and 'Über quantentheoretische Kinematik und Mechanik', *Mathematische Annalen* **95**, 1926, pp. 683–705.

[32]Cp. P. A. M. Dirac, *The Principles of Quantum Mechanics*, fourth edition, 1958, pp. 1 ff.

[33]M. Born, 'Bemerkungen zur statistischen Deutung der Quantenmechanik', in F. Bopp (ed.), *Werner Heisenberg und die Physik unserer Zeit*, 1961, pp. 103–118; cp. p. 104.

was suggested by a remark of Einstein's about the connection between the wave theory of light and the photon hypothesis. The intensity [of course, what is meant is the square of the amplitude] of the light waves was to be a measure of the density of the photons or, more precisely, of the probability of photons being present.'

Thus, owing to Born's statistical interpretation of matter waves, even the one problem of quantum theory which appeared not to be statistical—the problem of atomic stability—was reduced to, or replaced by, a statistical problem: Bohr's quantized 'preferred orbits' turned out to be those for which the *probability* of an electron's presence differed significantly from zero.

All this is to support my thesis that *the problems of the new quantum theory were essentially of a statistical or probabilistic character.*

2. My second thesis is that *statistical questions demand, essentially, statistical answers.* Thus quantum mechanics must be, essentially, a statistical theory.

I believe that this argument (although its validity is by no means generally admitted) is perfectly straightforward and logically cogent. (The argument may be traced back to Richard von Mises[34] and to my *L.Sc.D.*; it has been beautifully illustrated by Alfred Landé.[35])

Statistical conclusions cannot be obtained without statistical premises. And therefore answers to statistical questions cannot be obtained without a statistical theory.

Yet largely owing to the fact that the *problems* of the theory were not (and still often are not) seen to be statistical, other reasons were invented to explain the widely admitted statistical character of the theory.

Foremost among those reasons is the argument that it is our (necessary) *lack of knowledge*—especially the limitations to our knowledge discovered by Heisenberg and formulated in his '*principle of indeterminacy*' or '*principle of uncertainty*'—which forces us to adopt a probabilistic, and consequently a statistical, theory. (This argument is criticized in my fifth thesis below.)

[34]R. von Mises, *Probability, Statistics and Truth*, 1939 (first German edition 1928; third, 1951).
[35]A Landé, *Foundations of Quantum Theory*, 1955, pp. 3 f.; *New Foundations of Quantum Mechanics*, 1965, pp. 27 ff. and 39.

3. My third thesis is that it is this mistaken belief that we have to explain the probabilistic character of quantum theory by our (allegedly necessary) *lack of knowledge,* rather than by the statistical character of our problems, which has led to *the intrusion of the observer, or the subject, into quantum theory.* It has led to this intrusion because the view that a probabilistic theory is the result of lack of knowledge leads inescapably to the *subjectivist interpretation of probability theory;* that is, to the view that the probability of an event measures the degree of somebody's (incomplete) knowledge of that event, or of his 'belief' in it.

However, as I have tried to show for many years, it would be sheer magic if we were able to obtain knowledge—statistical knowledge—out of ignorance.[36]

4. My fourth thesis is that, as a consequence, we are faced with what I shall call *the great quantum muddle.* (It seems to me that the only adherent to Bohr's 'Principle of Complementarity' who is free of this muddle—following almost exactly Einstein's despised ideas in a new garb—is Fritz Bopp.[37])

In order to explain this great muddle, I shall have to say a few words about statistical theories.

Every probabilistic or statistical theory assumes the following:

(a) Certain *events* (5 turning up) which happen to certain *elements* (dice) in certain *experimental situations* (being shaken in a beaker, and thrown on a table). These form the 'population' for our statistics.

(b) Certain physical properties of these events, elements, and experimental situations; for example that the dice are of homogeneous material, and that only one of the six sides is marked with a '5'; and that the experimental situation permits a certain width of variation.

(c) A set of the *possible* events (*possible* under the experimental

[36]Cp. my *Logic of Scientific Discovery,* 1959 (first German edition 1934, second 1966); 'Probability Magic, or Knowledge out of Ignorance', *Dialectica* 11, 1957, pp. 354–74; 'The Propensity Interpretation of Probability' *British Journal for the Philosophy of Science* 10, 1959–60, pp. 25–42; and 'The Propensity Interpretation of the Calculus of Probability, and the Quantum Theory', in S. Körner, ed.: *Observation and Interpretation in the Philosophy of Physics,* 1962, pp. 65–70, 88–89. See also *Realism and the Aim of Science,* Vol. I of this *Postscript,* especially Part II.

[37]Cp. his paper cited in footnote 10 above.

conditions), called the points in the *sample space* or the *probability space* (the notion stems from Richard von Mises).

(d) A number associated with each point (or, in the case of a continuous sample space, with each region) of the sample space, determined by some mathematical function, called the distribution function. (The sum of these numbers is equal to 1; this can be achieved by some 'normalization'.) In the continuous case the distribution function is a density function.

Example: our sample space may be the United Kingdom, or more precisely, the set of events of a man or a woman living at some spot in the United Kingdom. The distribution function can be given by a (continuous) density distribution (normalized to 1) of the population; that is, the actual number of people living in a region, 'normalized' by being divided by the total population of the United Kingdom. We then can say that this *information* helps us to answer all questions of the type: what is the probability that an Englishman lives at a certain spot (region); or that an Englishman lives in 'the South of England'? (Here we assume that we have a proper division between North and South.)

Now it is clear that the statistical distribution function (whether normalized or not) may be looked upon as *a property characterizing the sample space*—in our case the United Kingdom. It is *not* a physical property characteristic of the *events* (5 turning up; or of Mr Henry Smith—a resident in the United Kingdom—being domiciled in Oxford); still less is it a property of the *elements* (the die; or Mr Smith).

This is particularly clear of Mr Smith: he is, for the statistical theory, nothing but an element under consideration. (In fact, the statistical theory will tell us almost the same about Mr Smith as it tells us, say, about his bed or his wristwatch: the statistical distribution of these physically very different elements will be almost identical). It is, perhaps, less clear of the die: in this case the distribution function is, we conjecture, *related* to its physical properties (its having six sides, the homogeneity of its material). However, this relation is not as close as it may seem at first sight. For the distribution function will be the same for big or small dice, and for dice made of some light plastic or of uranium. And the probability of 5 turning up will be the same for all dice that have only one side marked '5'—whatever the markings of the other sides may

be (though these may greatly influence other probabilities); and it will be a different one for all dice having more, or fewer, than one side marked '5', or for non-homogeneous dice.

Now what I call the great quantum muddle consists in taking a distribution function, i.e., a statistical measure function characterizing some *sample space* (or perhaps some 'population' of events), and treating it as *a physical property of the elements of the population*. It *is* a muddle: the sample space has hardly anything to do with the elements. There is no symmetrical relationship, and thus no 'duality', between particles and waves, or between particles and their associated field.

Unfortunately, many people, including physicists, talk as if the distribution function (or its mathematical form) were a property of the *elements* of the population under consideration. They do not discriminate between utterly different categories or types of things, and rely on the very unsafe assumption that 'my' probability of living in the South of England is, like 'my' age, one of 'my' properties—perhaps one of my physical properties.

Now my thesis is that this muddle is widely prevalent in quantum theory, as is shown by those who speak of a 'duality of particle and wave' or of 'wavicles'.

For the so-called 'wave'—the ψ-function—may be identified with the mathematical form of a function, $f(P, dP/dt)$, which is a *function of a probabilistic distribution function P*, where $f=\psi=\psi(q,t)$, and $P=|\psi|^2$ is a density distribution function. (See for example footnote 6, with a reference to E. Feenberg, in H. Mehlberg's excellent discussion of Landé's views.[38]) On the other hand, the *element* in question has the properties of a particle. The wave shape (in configuration space) of the ψ-function would be from this point of view a kind of accident which poses a problem to probability theory, but which has next to nothing to do with the physical properties of the particles. It is as if I were called a 'Gauss-man' or a 'non-Gauss-man' in order to indicate that the distribution function of my living in the South of England has a Gaussian or non-Gaussian shape (in an appropriate sample space).

[38][H. Mehlberg, 'Comments on Landé's "From Duality to Unity in Quantum Mechanics" ', in H. Feigl and G. Maxwell (eds), *Current Issues in the Philosophy of Science*, 1961, pp. 360–70.

5. My fifth thesis concerns Heisenberg's famous formulae:[38a]

$$\Delta E \Delta t \gtrsim h, \tag{1}$$

$$\Delta p_x \Delta q_x \gtrsim h. \tag{2}$$

These formulae are, beyond all doubt, validly derivable *statistical formulae* of the quantum theory. But they have been *habitually misinterpreted* by those quantum theorists who said that these formulae can be interpreted as determining some upper limits to the

[38][*(Added 1981) It is now often asserted that the first of these relations,

$$\Delta E \Delta t \approx h$$

is incorrect, invalid, or what not, because t is not an operator; and it has even been asserted that 'Δt' is a symbol without meaning (as if I could not open the shutter of my camera for the period $\Delta t = 1/100$ seconds; or emit a light signal of the same length of time). Even the purely formal argument that time is not an operator is no longer valid. (*Cp.* I. Prigogine, *From Being to Becoming*, 1980, pp. 206–231.) But there is altogether more to this.

Heisenberg's relations (the indeterminacy or uncertainty relations or, as I interpret them, the 'statistical scatter relations' (see *L.Sc.D.*, pp. 215, 225f.) are indeed part of the formalism of quantum mechanics; but in the arguments of Heisenberg and Bohr, and especially also in their arguments with Einstein as reported by Heisenberg in *Physics and Beyond*, 1971, pp. 80f., and by Bohr in P. A. Schilpp, ed., *Albert Einstein: Philosopher-Scientist*, 1949, pp. 199–241, the derivation of the Heisenberg formulae from the axioms of quantum mechanics plays no part whatever (not even their logical relation to the commutation relations being discussed). What is discussed are arguments from classical physics, especially wave optics, which are used to show, for example, that *classical* physics does not allow us to measure a particle's position and momentum, or the energy (frequency) of a beam of light, and its temporal length Δt, more precisely than Heisenberg's formulae permit. In other words, these formulae are not defended by showing that they are forced upon us by quantum mechanics, and that they are derivable from quantum mechanics, but they are defended as consequences of classical physics.

The only deviation from this is that use is sometimes made of the 'particle wave duality'. This is used essentially in the pre-quantum mechanical form: it is assumed that light consists of photons whose density (statistical probability) is determined by the optical wave amplitudes; or use is made of Born's interpretation of Schrödinger's equation.

This use of the particle-wave duality (which Einstein of course accepted) is sometimes needed for making the classical argument convincing. I suppose that this is one of the reasons why Bohr laid so much stress on this duality.

What is the reason behind this treatment of the Heisenberg formulae? I suppose it is an attempt at showing the reality of the indeterminacy relations (only classical physics treats physical reality) and thus the need for quantum mechanics: it is a justification of quantum mechanics as unavoidable even from the classical, realistic point of view. Moreover, it is an attempt to defend the *necessarily probabilistic* character of atomic theory with classical means (and independently of the non-commutative matrix algebra or operator algebra).]

precision of our measurements (or some lower limits to their imprecision).

My thesis is that these formulae set some lower limits to the *statistical dispersion or 'scatter'* of the results of sequences of experiments: they are *statistical scatter relations*. They thereby limit the precision of certain individual *predictions*.

But I also assert that *in order to test these scatter relations, we have to be able (and are able) to make measurements which are far more precise than the range or width of the scatter.*

The situation is like this: a statistical theory may tell us something about the distribution or scatter of the population in the environment of industrial towns. In order to *test* it, it will be necessary to fix the places where people live with a precision far exceeding the range of the predicted scatter. Our statistical laws may tell us that we cannot reduce the scatter below a certain limit. But to conclude from this that we are unable to 'measure' the positions of the places where the people live more precisely than the minimum statistical scatter is a mistake.

Since Heisenberg's formulae in their various proper interpretations are (as will be shown in detail in my next thesis) *statistical laws of nature, derivable from a statistical theory,* it is quite obvious that it is impossible to use them in order to explain why quantum mechanics is probabilistic or statistical. Moreover, being statistical laws, they *add* to our knowledge: it is a mistake to think that they set limits to our knowledge. What they do set limits to is the scatter of particles (or more precisely, the scatter of the result of sequences of certain experiments with particles). This scatter, they tell us, cannot be suppressed. It is also a mistake to think that the alleged limitation to our knowledge could ever be validly used for explaining the statistical character of the quantum theory. (See my eighth thesis, below.) And ultimately, it is the same mistake again if it is said that the Heisenberg formulae provide us with that *vagueness* which is allegedly needed for asserting without inconsistency the 'dual' character of particles and waves; that is, their character as 'wavicles'.

6. My sixth thesis is that, however remarkable the statistical laws of the theory are, including the Heisenberg formulae (1) and (2), they refer to a population of *particles* (or of experiments with *particles*) which are, quite properly, endowed with positions and

momenta (and mass-energy, and various other physical properties such as spin). It is true that the scatter relations tell us that *we cannot prepare experiments* such that we can avoid, upon repetition of the experiment, (1) scattering of the energy if we *arrange* for a narrow time limit, and (2) scattering of the momentum if we *arrange* for a narrowly limited position. But this means only that there are limits to the *statistical homogeneity* of our experimental results. Yet not only is it possible to *measure* energy *and* time, or momentum *and* position, with a precision greater than formulae (1) and (2) seem to permit, but *these measurements are necessary for testing the scatter* predicted by these very formulae.

I shall now try to produce some arguments for what I have said in my last two theses. These arguments will show, incidentally, that the Heisenberg formulae (1) and (2) can be derived from theories which are much older than the commutation relations of quantum mechanics.[39]

We can derive Heisenberg's formula

$$\Delta E \, \Delta t \geqq h$$

(1)

from Planck's quantum condition of 1900

$$E = h\nu.$$

This leads, in view of the constancy of h, at once to

$$\Delta E = h \, \Delta\nu,$$

a formula in which 'Δ' may be interpreted in various ways. In order to obtain Heisenberg's formula (1) we only have to combine the last formula with a principle of optics, the *principle of harmonic resolving power*, which is even older than Planck's condition. (Both Heisenberg and Bohr base their derivations of the indeterminacy relations directly or indirectly upon this principle.[40]) This principle states that if a monochromatic wave train of frequency ν is cut up by a *time shutter* into one stretch or several stretches ('wave packets') of the duration Δt, then the width $\Delta\nu$ of the spectral line will become

$$\Delta\nu \geqq 1/\Delta t.$$

[39][See *Objective Knowledge*, 1972, pp. 301–4. Ed.]

[40]Cp. W. Heisenberg, *The Physical Principles of the Quantum Theory*, 1930, pp. 21 and 27.

This is, for various reasons, a remarkable law. (It contains the principle of superposition.) It leads from

$$\Delta E = h\,\Delta\nu$$

immediately to

$$\Delta E \geqq h\,/\,\Delta t$$

and thus to formula (1).

But in so deriving formula (1), *we are no longer free* to interpret 'Δ' in various ways (for example, as the width of imprecision of a measurement). We are, rather, bound in our interpretation by the meaning given to 'Δ' by the principle of harmonic resolving power. This principle interprets '$\Delta\nu$' as the width of spectral lines. Accordingly, Planck's principle (in Einstein's interpretation) forces us to interpret this width as the scatter of the energy of the particles (photons) which make up the spectral lines; for a spectral line of frequency ν is to be interpreted as the statistical result of incoming photons of energy $E=h\nu$, and consequently the width $\Delta\nu$ of the spectral line as the range ΔE of the *statistical scatter* of the energies of the photons which together form the spectral line. Thus formula (1) states the law that, if we vary at will the period Δt of our shutter, we are bound to influence inversely the scatter ΔE of the energy of the incoming photons.

This derivation shows clearly that (1) is a *statistical law,* and part of the statistical theory. It can be tested only by ascertaining the distribution of the incoming photons on the photographic film or plate; and in order to do this, we must measure the places where the photons hit the spectral line with an imprecision, say δE, very much smaller than the width ΔE of the line:

$$\delta E \ll \Delta E.$$

Thus the testing of the law expressed by (1) and of its statistical predictions demands that we can *measure* the incoming particle with a precision δE which satisfies

$$\delta E\,\Delta t \ll h.$$

This kind of thing is done every day; and it shows that the

Heisenberg formulae are valid for *statistical predictions about many particles, or about sequences of many experiments* with individual particles, but that they are misinterpreted as limiting the precision of *measurements* of individual particles.

There is a derivation of the second Heisenberg formula

$$\Delta p_x \Delta q_x \gtrsim h \tag{2}$$

which is analogous to the derivation with the help of the *time shutter*. We start again with a (flat) monochromatic wave train v and cut it; this time by a *screen* (vertical to the direction z of the beam), *with one slit* of variable width Δq_x (the '*one-slit experiment*'). When the slit is very wide, there will be only a marginal effect upon the wave train. But when it narrows, we get a scattering (diffraction) effect: the narrower the slit Δq_x, the wider will be the angle by which the rays diverge from their original direction: here another form of the principle of harmonic resolving power applies (\bar{v}_x is the projection on the x-axis of the wave number, that is, the number of waves per centimetre):

$$\Delta \bar{v}_x \approx 1/\Delta q_x.$$

Multiplying both sides by h we get

$$h \Delta \bar{v}_x \approx h/\Delta q_x.$$

Using instead of Planck's formula $E = hv$ that of de Broglie in the form $p_x = h\bar{v}_x$, we can write 'Δp_x' for $h\Delta\bar{v}_x$; and so we arrive at (2).

When the slit Δq_x is very small we obtain, according to Huygens's principle, waves emerging from it which spread not only in the z direction but also in the $+x$ and $-x$ direction (cylinder waves). This means that the particles which, before reaching the slit, had a momentum $p_x = 0$ (since they were proceeding in the z direction), will now have a considerable scatter of momenta Δp_x, in the $+x$ and $-x$ direction. We can test this scatter again by measuring the various momenta with a spectrograph in various positions. There is, in principle, hardly a limit to the precision δp of the measurements of the various momenta in the various directions; that is, we have again

$$\delta p_x \ll \Delta p_x$$

and thus

$$\delta p_x \Delta q_x \ll h.$$

57

Again, we could not test the statistical law (2) in this *one-slit experiment* without these more precise measurements $\delta p_x \ll \Delta p_x$.

Incidentally, *we measure the momentum, p_x,* of the incoming particle *by its position* on the film of the spectrograph. And this is typical. It should hardly be necessary to stress that we almost always measure momenta by positions. (For example, if we measure a Doppler effect, we do so with the help of a spectral line, that is by measuring the position of the line on a photographic plate.) It has, unfortunately, become necessary to emphasize this point, because of Bohr's repeated assertion that position measurements and momentum measurements are incompatible (and 'complementary') owing to 'the mutual exclusion of any two experimental procedures, permitting the unambiguous definition of complementary physical qualities . . . '.[41] The two experimental procedures, we are told by Bohr, exclude each other because the momentum measurements need a *movable screen*,[42] while the position measurement needs a *fixed screen* or a fixed photographic plate. But we often measure momenta by fixed photographic plates, that is, by positions; but never by a movable screen. (Incidentally, the use of Bohr's movable screen would entail at least two *position measurements* of the screen.)

A famous problem for particle theory is posed by the *two-slit experiment* (or *n*-slit experiment), with two (or more) slits with the (periodic) distance Δq_x between them. Alfred Landé has suggested that the situation with regard to the two-slit experiment could be explained with the help of what one may call the Duane-Landé space-periodicity formula:[43]

$$\Delta p_x = n\,h\,/\,\Delta q_x \qquad\qquad (n = 1, 2, \ldots)$$

This formula, which is due to a period prior to quantum mechanics, is of course in its turn derivable from quantum mechanics.

[41]N. Bohr: 'Discussion with Einstein on Epistemological Problems in Atomic Physics', in P. A. Schilpp (ed.), *Albert Einstein: Philosopher-Scientist,* 1949, pp. 201–31; *cp.* p. 234 where he quotes from his reply ('Can Quantum-Mechanical Description of Physical Reality be Considered Complete?', *Physical Review* **48**, 1935, pp. 696–702) to the paper by Einstein, Podolsky and Rosen cited in footnote 14 above.

[42]As depicted in 'Discussion with Einstein on Epistemological Problems in Atomic Physics', p. 220.

[43]A. Landé: *New Foundations of Quantum Mechanics,* 1965, pp. 9–12.

The two slit experiment turns out to be a space-periodicity experiment with the periodicity Δq_x. The particles transfer to the screen (or the grid) a momentum packet Δp, or its multiples, such that

$$\Delta p_x \Delta q_x = h.$$

As a consequence (as shown by Landé, *loc. cit.*) we get the wave-like fringes.

The usual question 'how does the particle which goes through slit 1 "know" that slit 2 is open rather than closed?' can now be reasonably well cleared up. It is the *screen* (or the grid, or the crystal) which 'knows' whether there is a periodicity Δq_x built into it or not, and which therefore 'knows' whether it can absorb momentum packets of the size $\Delta p_x = h / \Delta q_x$. The particle does not need to 'know' anything: it simply interacts with the screen (which 'knows'), according to the laws of conservation of momentum *and* of space periodicity; or more precisely, it interacts with the total experimental arrangements (see my eighth and especially my tenth theses below). [Incidentally, E. L. Hill suggested a similar solution of this riddle in a lecture at the Minnesota Center for Philosophy of Science in the autumn of 1962 which I had the good fortune to attend.]

I have so far spoken mainly about particles and their (indirect) measurements: for example, momentum measurements by way of position measurements. But there are other methods, of course: Geiger counters may measure (not very precisely) position, *and* time; and so may Wilson chambers. And the position measurement in a Wilson chamber may be an indirect momentum measurement. However, the time measurement of the incoming particle may be of particular interest to us in every case in which the frequency (or energy) of the emission is very sharp—as it is in the classical case of a Bohr hydrogen atom.

Here we have Rydberg's constant R, a wave number, so that Rc is a constant frequency, ν_R, which can be calculated, according to Bohr (1913), with great precision from the constants of the theory (μ is the mass of the electron, e its charge):

$$\nu_R = Rc = 2\pi^2 e^4 \mu / h^3.$$

Then the Rydberg-Ritz combination principle (formulated by Ritz, using Rydberg's constant, in 1908, five years before Bohr's theory

of the hydrogen atom) asserts for the frequencies, $\nu_{m,n}$, of emission or absorption the relation

$$\nu_{m,n} = \nu_R/m^2 - \nu_R/n^2 \qquad (m,n = 1,2, \ldots)$$

Multiplied by h this becomes Bohr's quantization rule of emission and absorption (1913); [partly anticipated by Arthur Haas (who based his calculation on the model of the atom) in 1910.] Thus the permissible frequencies $\nu_{m,n}$ and the various corresponding Bohr-energies of the particles can be calculated from first principles, as it were—they are variables which can take on only certain discrete values ('*eigenvalues*', which might be described as quasi-constants). Accordingly, $\Delta\nu_{m,n}$ may be extremely small, and Δt, calculated with the help of the principle of harmonic resolving power, will be large.

But these sharp spectral lines (which would lose their sharpness if interfered with by a time shutter) can be statistically investigated by timing the arrivals of the photons (which gives also the time of emission) by means such as a Wilson chamber or a Geiger counter. (Especially impressive here are the Compton-Simon photographs of high frequency X-ray photons of very precise frequency or energy.) For these arrival times we may get $\delta t \ll \Delta t$, and thus

$$\Delta E \; \delta t \; \ll h.$$

7. My seventh thesis is that all this, or most of it, was in effect admitted by Heisenberg (and, incidentally, also by Schrödinger).

First I would repeat that the *predictions* of the theory are statistical, with a scatter given by the Heisenberg formulae. The *measurements* which must be *more precise than the scatter* (as I have pointed out) may serve as *tests* of these predictions: *these measurements are retrodictions.*

Heisenberg saw, and said, that these highly precise retrodictive measurements were possible. *What he did not see was that they had a function in the theory—that they were needed for testing it* (and that they could be tested in their turn).

Thus he suggested, half-heartedly and yet pretty strongly (see the text to n. 47 below), that these retrodictive measurements were *meaningless.* And this suggestion was taken up and *turned into a dogma* by the adherents of the Copenhagen interpretation, especially when it was found that there were no vectors in Hilbert space

corresponding to any measurements sharper than the formulae (1) and (2).

But this fact does not really create any difficulty. *The vectors in Hilbert space correspond to the statistical assertions of the statistical theory.* They say nothing about measurements, or about the *tests* of the statistical assertions by the determination of the position and momentum, or of the energy and time, of individual particles.

I shall now quote some evidence for my thesis regarding Heisenberg's admission that the measurements I have described can be made, and his suggestion that they are, if not completely meaningless, at best pointless and uninteresting, because they merely refer to the past. He says that measurements which 'can never be used as initial conditions in any calculation of the future progress of the electron and [which] thus cannot be subjected to experimental verification' are devoid of physical significance.[44] But this is a double mistake. For (a) the preparation of initial conditions admittedly is very important, but so are test statements which *always* look into the past and whose *main* function is not to be 'verifiable' (that is, testable) in their turn but to 'verify' (or more precisely to test). And (b) it is a mistake to think that these test statements, though looking into the past, are not 'verifiable' (or more precisely, testable) in their turn. On the contrary, it is one of the principles of the quantum theory that every measurement can be 'verified' or tested in the sense that its immediate repetition will yield the same result. (This principle, whose author seems to be von Neumann, is not generally valid, unless it is trivially so in the sense explained below, under the heading of my ninth thesis.) Thus to say that these measurements which look into the past 'cannot be subjected to experimental verification' is simply mistaken.

In order to show quite definitely that Heisenberg and I are talking about the same measurements, and that we are in agreement that they are not subject to the uncertainty relations, I wish to remind the reader, as explained above, of the fact that the measurements of p_x with the help of spectrographs at various positions (or of a photographic plate parallel to the horizontal screen) are, in fact, *position* measurements, so that we obtain our total information

[44]*Cp.* W. Heisenberg, *The Physical Principles of the Quantum Theory*, 1930, p. 20.

about position and momentum by way of *two position measurements*: the first is provided by the slit Δq_x, the second by the impact of the particle on the photographic plate. (We can take the frequency—or energy—of the beam as known.) Now it is precisely about such an arrangement consisting of *two position measurements* (which allow us to calculate the position *and* momentum *after* the first and *before* the second measurement) that Heisenberg says the following:[45] 'The . . . most fundamental method of measuring velocity [or momentum] depends on the determination of position at two different times . . . it is possible to determine *with any desired degree of accuracy* the velocity [or momentum] before the second [measurement] was made; but it is the velocity *after* this measurement which alone is of importance to the physicist . . .'

Heisenberg is even more emphatic concerning experiments in which we measure the position of a particle whose momentum is known (say, because the particle belongs to a monochromatic beam): '. . . the uncertainty relation does not refer to the past', he writes;[46] 'if the velocity of the electron is at first known and the position then exactly measured, the position for times previous to the measurement may be calculated. Then for these past times $\Delta p \Delta q$ is smaller than the usual limiting value.' So far we can agree. But now comes our subtle but important disagreement; for Heisenberg continues: 'but this knowledge of the past is of a purely speculative character, since it can never . . . be used as an initial condition in any calculation of the future progress of the electron' (this I believe to be true) 'and thus cannot be subjected to experimental verification' (this is false, as I shall show).

Heisenberg adds to this: 'It is a matter of personal belief whether such a calculation concerning the past history of the electron can be ascribed any physical reality or not.'[47]

Almost every physicist who read Heisenberg opted for 'not'.

But it is not a matter of personal belief: the measurements in question are *needed* for testing the statistical laws (1) and (2); that is, the scatter relations.

The particular case in question, of a position measurement of a

[45]Cp. W. Heisenberg, *op.cit.*, p. 25; the italics are mine and I have changed the position of a phrase to improve the readability by avoiding an ambiguity.
[46]Cp. W. Heisenberg, *op.cit.*, p. 26.
[47]*Loc. cit.*

particle from which retrodictively 'the positions for times previous to the measurement may be calculated', as Heisenberg puts it, plays a *most* important role in physics: if we measure the position of a particle (a photon or an electron) on the photographic film of any spectrograph, then we use this position measurement (together with the known arrangement of the experiment) for calculating, with the help of the theory, the frequency or energy and thus the momentum of the particle; always, of course, retrodictively. To question whether the so ascertained 'past history of the electron can be ascribed any physical reality or not' is to question the significance of an indispensable standard method of measurement (retrodictive, of course); indispensable, especially, for quantum physics.

But once we ascribe physical reality to measurements for which, as Heisenberg admits, $\Delta p \Delta q \ll h$, the whole situation changes completely: for now there can be no question whether, according to the quantum theory, *an electron can 'have' a precise position and momentum. It can.*

But it was just this fact that was constantly denied: although Heisenberg made it 'a matter of personal belief', Bohr and the Copenhagen interpretation (partly because of the non-existence of those vectors in Hilbert space) *insisted that an electron just cannot have a sharp position and momentum at the same time.* This dogma is the core of Bohr's thesis that quantum theory is 'complete', presumably in the sense that a particle cannot have properties which the theory (allegedly) does not allow to be 'measured'.

Thus the so-called 'paradox' of Einstein, Podolsky, and Rosen (see n. 14 above) is not a paradox but a valid argument, for it established just this: that we must ascribe to particles a precise position *and* momentum, which was denied by Bohr and his school (though it is admitted by Bopp).

The Einstein-Podolsky-Rosen thought experiment has since become a *real* experiment, in connection with particle pair-creation, and particle pair-destruction creating a photon-pair. The times and energies of the pairs can be in principle measured with any degree of precision. Of course, the measurements are *retrodictive:* they are *tests* of the theory.[48]

[48]See, for example, O. R. Frisch, 'Observation and the Quantum', in M. Bunge (ed.), *The Critical Approach to Science and Philosophy*, 1964, pp. 309–15. [See also the 'Preface 1982' to this volume. Ed.]

Why did Bohr and his followers deny that $\delta p_x \delta q_x \ll h$ is possible? Because of the great quantum muddle, the alleged dualism of particle and wave: it is said that there are two *'pictures'*, the *particle picture* and the *wave picture*, and that they have been shown to be equivalent or 'complementary'; that is to say, both valid. But this 'complementarity' or 'duality' must break down, it is said, if we allow the particle to have at the same time a sharp position and a sharp momentum.

It is from here, and from the subjective interpretation of probability to which we shall turn next, that the subjectivist interpretation of the quantum theory arose—almost of necessity.

8. My eighth thesis results from an attempt of mine to explain, though not to excuse, the great quantum muddle, as I have called it. My thesis is that *the interpretation of the formalism of quantum mechanics is closely related to the interpretation of the calculus of probability.*

By the calculus of probability I mean a *formal calculus* which contains *formal laws* such as

$$0 \leqq p(a,b) \leqq 1$$

where '$p(a,b)$' may be read 'the probability of a relative to b' (or 'the probability of a given b').

What 'probability' (the function or functor 'p') *means*, and what the arguments 'a' and 'b' *stand for*, is left open to *interpretation*.

It is assumed, however, that there is a set of entities, S, say, to which the variables a,b,c, \ldots, refer; and that if a belongs to S, then $- a$ (read 'non-a') also belongs to S; and that if a and b belong to S, then ab (read 'a-and-b') also does. No further assumption needs to be made about S. (More especially, we do not need to assume that S is a partially ordered set.) However, it is assumed that the range of admissible meanings of all these symbols, though open to many different interpretations, is somewhat confined by a number of formal rules which connect these symbols, rules which pertain to the p-function.[49]

The following formulae are trivial examples of such formal rules:

[49]Cp. my *Logic of Scientific Discovery*, 1959, New Appendices *iv and *v. [See also *Realism and the Aim of Science*, Vol. 1 of this *Postscript*, Part 2. Ed.]

$$p(a,a) = 1$$
$$p(a,b)+p(-a,b) = 1 \text{ (unless } b \text{ is such that } p(-b,b) \neq 0).\text{[50]}$$
$$p(a,b) = p(aa,b) = p(a,bb).$$
$$p(a,c) \geqq p(ab,c) \leqq p(b,c).$$
$$p(ab,c) = p(ba,c).$$
$$p((ab)c,d) = p(a(bc),d).$$
$$p(a,b) = p(a,c), \text{ whenever } p(a,a) = p(b,c) = p(c,b).$$

We may also give a definition of 'absolute probability', $p(a)$, in terms of 'relative probability', $p(a,b)$:

$$p(a) = p(a,-((-a)a)).$$

(The definition in the opposite direction is more widely known: provided $p(b) \neq 0$, $p(a,b) = p(ab)/p(b)$.)

The task of selecting a number of these *formal rules* so that all the others are derivable from them, is the task of finding one or more suitable *axiomatizations* of the formal calculus of probability.[51] I mention it only in order to contrast it with the task of finding one or more suitable *interpretations*.[52]

There is a great variety of interpretations, which may be divided into two main groups: the *subjective* and the *objective* interpretations.

The *subjective interpretations* are those which interpret the number $p(a,b)$ as measuring something like our knowledge, or our belief, in (the assertion) a, given (the information) b. Thus the arguments of the p-function, that is, a,b,c, \ldots are in this case to be interpreted as items of belief or doubt, or items of information, or propositions, or assertions, or statements, or hypotheses.

For a long time it was thought (and is still thought by many eminent mathematicians and physicists)[53] that we may start from a

[50]This somewhat unusual condition is explained and discussed at length in my formal axiomatization of the probability calculus in *L.Sc.D.*, New Appendices *iv and *v. All the formal rules given here in connection with the eighth thesis are theorems which are derivable from my axioms.

[51]*Cp.* my *Logic of Scientific Discovery*, and 'Creative and Non-Creative Definitions in the Calculus of Probability', *Synthese* 15, 1963, pp. 167–86; see also corrections in *Synthese* 21, 1970, p. 107.

[52]*Cp.* my 'Probability Magic, or Knowledge out of Ignorance', *Dialectica* 11, 1957, pp. 354–74.

[53]The great historical significance of this belief is discussed in Section II of the 'Preface 1982' to this volume.

subjectively interpreted system of probabilistic premisses *and then derive from these subjectivist premisses objective statistical conclusions. This is, however, a grave logical blunder.*

The blunder [or rather, the prehistory of the blunder, and not, as B. L. van der Waerden tells me[54], the blunder itself] may be traced back to some of the great founders of probability theory, to Jacob Bernoulli and especially Siméon Denis Poisson, who thought that they had discovered, in their derivations of the various forms of the *law of great numbers*, a kind of logico-mathematical *bridge* leading from non-statistical assumptions to statistical conclusions; that is, to conclusions concerning the *frequency* of certain events.

The logical mistake was carefully analysed by Richard von Mises[55] and also by myself.[56] Mises showed that at some stage or other in the derivation the non-statistical meaning of the symbols is dropped and tacitly replaced by a statistical one. This is usually done by interpreting a probability approaching 1 as 'almost certain' in the sense of 'almost always to happen', instead of 'almost certain' in the

[54][* (Added 1980) I wish to thank Professor van der Waerden for a most friendly letter in which he expressed his agreement with the present essay, with the exception of my historical assertion about the blunder made by the founders of probability theory. I must make it quite clear that in the field of the history of science I am an amateur (as in all other fields, I am afraid). But the following seems to me pretty clear.

The calculus of probability was interpreted by many of its founders (if not by all) as a theory for which there is a need and an application (and all applications of probability are statistical) arising from our *insufficient knowledge*, rather than from a lack of determination in objective nature. The famous statement of Laplace, in which he formulates determinism by calling in a 'vast intelligence' (later called the 'Laplacean demon'), at the beginning of his *Philosophical Essay on Probabilities*, has precisely the function of excluding any objective interpretation of probability and thus of asserting the subjective character of probability theory. The tradition thus established dominated physics at least until Heisenberg's indeterminacy relations (1927); and even these allowed a curious mixture of subjective and objective interpretations to survive. The main opponents defended a frequency interpretation (like the logician John Venn and the mathematician Richard von Mises).

Yet, even if my understanding of the historical origins of this blunder should turn out to be incorrect, that would not affect my main argument: that there *is* a logical blunder in trying to derive from subjectivist premisses *objective* statistical conclusions.]

[55]R. von Mises, *Probability, Statistics, and Truth*, 1939 (first German edition 1928; third, 1951).

[56]*Cp.* my *Logic of Scientific Discovery*, section 62; and 'Probability Magic' (see footnote 52 above).

sense of 'very strongly believed in' or perhaps 'almost known'. Sometimes the mistake consists in replacing 'almost certainly known' by 'known almost certainly to occur'. However this may be, the mistake is very clear: from premises about degrees of belief we can never get a conclusion about the frequency of events.

It is strange that this idea that we can derive statistical conclusions from premises expressing uncertainty is still so strong among quantum theorists; for John von Neumann, one of the most influential among them, accepted in his famous book, *Mathematical Foundations of Quantum Mechanics*, the theory of probability of von Mises.[57] Yet von Neumann's praise of this theory does not seem to have induced quantum theorists to study carefully von Mises's arguments against the existence of a '*bridge*' from non-statistical premises to statistical conclusions.

I do not wish to imply that I accept von Mises's theory as a whole; but I believe that his criticism of the alleged '*bridge*' from non-statistical premises to statistical conclusions is unanswerable; and I do not even know of any serious attempt to refute it. Nevertheless, the subjective theory, under the name of 'Bayesian probability', is widely and uncritically accepted.

I now proceed to the *objective interpretations* of the probability calculus. I shall here distinguish between *three* such interpretations:

(a) *The classical interpretation* (de Moivre, Laplace) which takes $p(a,b)$ to be the proportion of equally possible cases compatible with the event b which are also favourable to the event a. For example, let a be the event 'at the next throw of this die 5 will turn up' and take b to be the assumption '6 will *not* turn up' (or 'only throws other than 6 will be considered as throws'); then $p(a,b) = 1/5$.

(b) *The frequency interpretation* or statistical interpretation (John Venn, George Helm, von Mises) which takes $p(a,b)$ as the relative frequency of the events a among the events b. This interpretation, which I developed by trying to remove some of its difficulties,[58] is one which I upheld for about twenty years (from approximately 1930 to 1950), though I always stressed the existence of other interpretations.[59]

[57]Cp. J. von Neumann, *Mathematical Foundations of Quantum Mechanics*, 1949, 1955, p. 298, footnote 156.
[58]See my *Logic of Scientific Discovery*, Ch. VIII and Appendix *vi.
[59]See *op. cit.*, section 48.

(c) *The propensity interpretation* which I developed from a criticism of my own form of the frequency interpretation and which may at the same time be regarded as a refinement of the classical interpretation.

I shall have to say a few things about each of the three objective interpretations.

In favour of (a), the classical interpretation, it may be said that it is used, almost as a matter of course and obviously with good reason, in situations where we conjecture that we have before us something like *'equally possible cases'*: we do not need to experiment with a regular polyhedron in order to conjecture that, if it is of homogeneous material and has n sides, the probability for any of these sides to turn up in any one throw will be $1/n$.

On the other hand, the classical interpretation has been severely criticized on several counts, of which I will mention only two. (1) As it stands it is inapplicable to anything like unequally possible cases such as playing with a loaded die. (2) It succumbs, like the subjective interpretation, to von Mises's criticism: there is no logical or mathematical *bridge* (such as the law of great numbers was supposed to be) which leads from premises about *possibilities* to statistical conclusions about *relative frequencies*. (Mises showed this in great detail for Poisson's derivation of his law of great numbers.) Nor does it make much sense to say of a ratio of many favourable to many possible cases that, even if it approaches 1, it tells us what is 'almost certainly' going to happen: obviously, there occurs here (as von Mises stressed) a shift of meaning in proceeding from possibilistic premises to a statistical conclusion.

As to (b), the frequency interpretation, I feel confident that I have succeeded in purging it of all those allegedly unsolved problems which some outstanding philosophers like William Kneale have seen in it.[60] Nevertheless,[61] I found that a further reform was needed, and I have tried to respond to this need.

Thus I come to (c), to the *propensity interpretation* of probability.

Nothing is further from my mind than an attempt to solve the pseudo-problem of giving a definition of the meaning of probability. It is obvious that the word 'probability' can be used perfectly

[60]*Cp.* W. C. Kneale, *Probability and Induction*, 1949. (See also footnote 52 above.)

[61]*Cp.* my *Realism and the Aim of Science*, Part 2.

properly and legitimately in dozens of senses, most of which, incidentally, deviate from those senses which agree with the rules of the formal calculus of probability.[62] I do not even wish to say that the propensity interpretation of probability is the best interpretation of the formal probability calculus. I only wish to say that it is the best interpretation known to me *for the application of the probability calculus to a certain type of 'repeatable experiment'*, in physics, more especially, and also, I suppose, in related fields such as experimental biology.

Whether the propensity interpretation is applicable to *all* cases of betting is also really unimportant to my argument. It has been claimed that it is not applicable to the betting on horses in horse racing. Should this turn out to be so, I should simply recommend that a different interpretation be applied to this case. I know nothing about horse racing; but it seems to me that what one might want to know in betting on a horse are such things as its staying power, speed (maximum and minimum speed, as it were, on earlier occasions), state of health, and similar things which together may be described as the propensity of that horse to do well in a race (as compared, of course, with its competitors). Were someone—prior to placing a bet—to attempt to get information from a stable, this is just the sort of information that he might want to obtain.

However this may be, the formal probability calculus clearly is applicable to a large class of 'games of chance'. But I am not trying to propose universally satisfactory meanings for the words 'probable' and 'probability', or even a universally applicable interpretation of the formal calculus. I am trying to propose an interpretation of the probability calculus which is not *ad hoc,* and which solves some of the problems of the *interpretation of quantum theory*.

I shall explain the propensity interpretation as a development of the classical interpretation. The latter, it will be remembered, explains $p(a,b)$—that is, the probability of a given b—as the proportion of those equally possible cases satisfying b which are favourable to the event a.

[62]For such senses, and very important ones, see especially G. L. S. Shackle, *Decision, Order and Time in Human Affairs,* 1961; and C. L. Hamblin, 'The Modal "Probably" ', *Mind* **68**, 1959, pp. 234–40. (Incidentally, I do not think that our subjective 'faith' or 'belief' in an explanatory theory ever agrees with the rules of the calculus of probability.)

I propose, as a first step, to omit the word 'equally' and to introduce 'weights', and thus to speak, instead of 'numbers of cases', of the 'sum of the weights of the cases'. And I propose, as a second step, to interpret these 'weights' of the possibilities (or of the possible cases) as *measures of the propensity, or tendency, of a possibility to realize itself upon repetition*.

Thus I arrive at the following:

The formulae $p(a,b)$, or 'the probability (or propensity) of a given b', may be interpreted as 'the 'sum of the weights of the possible cases that satisfy the condition b which are also favourable to a, divided by the sum of the weights of the possible cases that satisfy b'.

The main idea of this interpretation can also be put as follows: I propose to distinguish for the moment *probability statements* from *statistical statements* (or statements of relative frequency), and to look upon probability statements as statements about frequencies in *virtual* (finite) sequences of well-characterized experiments, and upon *statistical statements* as statements about frequencies in *actual* (finite) sequences of such experiments. In probability statements the 'weights' attached to the possibilities are measures of these (conjectural) virtual frequencies, *to be tested by actual statistical frequencies*.

To use an example: if we have a large die containing a piece of lead whose position is adjustable, we may *conjecture* (for reasons of symmetry) that the weights (that is, the propensities) of the six possibilities are *equal* as long as the centre of gravity is kept equidistant from the six sides, and that they become *unequal* if we shift the centre of gravity from this position. For example, we may increase the *weight of the possibility* of 6 turning up by moving the centre of gravity away from the side showing the figure '6'. And we may here interpret the word 'weight' to mean 'a measure of the propensity or tendency to turn up upon repetition of the experiment'. More precisely, we may agree to take as our measure of that propensity the (virtual) *relative frequency* with which the side turns up in a (virtual, and virtually infinite) sequence of repetitions of the experiment.

We may then test our conjecture by a sequence of actual repetitions of the experiment.

In proposing the propensity interpretation I propose to look upon *probability statements* as statements about some measure of a

property (a physical property, comparable to symmetry or asymmetry) of the *whole repeatable experimental arrangement*;[63] a measure, more precisely, of a *virtual frequency*; and I propose to look upon the corresponding *statistical statements* as statements about the corresponding *actual frequency*.

In this way we easily get over the objection raised by von Mises against the classical interpretation, simply by replacing mere *possibilities* by *propensities* which we interpret as tendencies to produce relative frequencies on repetition of similar conditions or circumstances.

Two further points are very important:

First, the probability distribution is taken to be a property of the

[63][*(Added 1980) Concerning the word 'experiments': my theory of propensities was always intended as a completely objective physical theory. The propensities refer to any physical situation. Propensities are then not properties of the particle but of the objective physical situation; for example, of an experimental situation. Of course the objective situation will *normally* be one that has arisen in the physical world without human interference, although it *may* be due to man, and perhaps even to a physicist who has built an apparatus. In the latter case we speak of 'experimental arrangement'. My somewhat accidental use of this phrase in the text here has led some of my critics to suppose that I speak of propensities *only* relative to such *man-made* experimental arrangements. And they have concluded from this that I do not succeed in my attempt to exorcize 'the observer'; also, that my theory is essentially equivalent to Bohr's. Thus Feyerabend and Bunge—and Max Jammer, following Feyerabend—have even attributed to me a stress (or 'insistence') on *experimental* set-up which makes it appear that I am a subjectivist. Jammer writes (*The Philosophy of Quantum Mechanics*, 1974, p. 450) that 'Popper's interpretation and his polemics against Bohr were severely criticized by Paul K. Feyerabend in a comprehensive vindication of the complementarity interpretation . . . For Popper's insistence that the experimental conditions of the whole physical setup determine the probability distribution is, in Feyerabend's view, precisely what Bohr had in mind when he used the notion of a "phenomenon" to include the account of the whole experimental arrangement'. (*Cp.* Feyerabend: 'On a Recent Critique of Complementarity', *Philosophy of Science* **35**, pp. 309–331 (1968); and **36**, pp. 82–105 (1969)). This is taking as the basis of one's criticism an almost accidental formulation, disregarding the context, and all I have written on propensities. It is clear from my earlier work and also this book (see in particular the 'Metaphysical Epilogue' to the present volume and Chapter III, section 18 below), that propensities, as I call them, are completely objective and dependent *not* on *our* set-up of experiments, but simply on the physical situation which, in certain cases, may be experimentally controlled. In so far as a man-made experimental set-up is an objective physical fact, an objective physical situation, it leads to propensities *just like any other physical situation*. (See also Theses 10 and 11 below.) It is perfectly true, by the way, that it is not here that I disagree with Bohr. But this does not mean that I do not disagree with him.]

single experiment, and a probabilistic law (or a probability statement) as relative to some rule specifying the conditions for accepting a situation or an experiment as a *repetition of another one.* [64] For example, in dicing, the minimum time taken in shaking the beaker may or may not form part of this rule, or of these conditions or specifications.

Secondly, we can look upon probability as a *real physical property of any concrete unique physical situation, and therefore also of the single physical experiment* or, more precisely, of the *experimental conditions* laid down by the rule that defines the conditions for the (virtual) *repetition* of the experiment.

A propensity is thus a somewhat abstract kind of physical property; nevertheless it is a real physical property. To use Landé's terminology, *it can be kicked, and it can kick back.*

Take, for example, an ordinary symmetrical pin board, so constructed that if we let a number of little balls roll down, they will (ideally) form a normal distribution curve. This curve will represent the *probability distribution* for each single experiment with each single ball of reaching a certain possible resting place.

Now let us 'kick' this board; say, by slightly lifting its left side. Then we also kick the propensities—the probability distribution— since it will become slightly more probable that any single ball will reach a point towards the right end of the bottom of the board. And the propensity will kick back: it will produce a differently shaped curve formed by the balls if we let them roll down and accumulate.

Or let us, instead, remove *one pin.* This will alter the probability for every single experiment with every single ball, *whether or not the ball actually comes near the place from which we removed the pin.* (This has its similarity with the two-slit experiment, even though we have here no superposition of amplitudes;[65] for we may ask: 'How

[64]In this sentence I have added the words: 'and a probabilistic law (or a probability statement) as', in order to emphasize the point that probability statements (in physics) are lawlike (since they are relative to virtual repetitions). I have made a similar change in the third paragraph of my tenth thesis below, where I have added the phrase 'Propensity statements in physics describe' in place of 'They [propensities] are'.

[65][This remark makes it clear that Feyerabend did not read this paper carefully before objecting (*op. cit.*) to Popper that the pin board probabilities are additive, whereas in quantum theory, where there is superposition of amplitudes, the situation is not additive. Below, in the ninth thesis, Popper also writes of the pinboard: 'There will be no interference of amplitudes: if we have two slits Δq_1 and

can the ball "know" that a pin has been removed if it never comes near the place?' The answer is: the ball does not 'know'; but the board as a whole 'knows', and changes the probability distribution, or the *propensity*, for *every* ball; a fact that can be tested by statistical tests.)

Thus we can 'kick' the probability field by making certain changes (either gradual or abrupt ones) in the conditions of the experiment, and the field 'kicks back' by changing the propensities, an effect which we can test statistically by repeating the experiment under the changed conditions.

But there are further important aspects of the propensity interpretation, which we can again illustrate with the help of the pin board.

Let us leave the pin board in its ordinary (symmetrical) state; we then can ask for the probability distribution of reaching the various final positions *for those balls which satisfy the condition of hitting a definite pin* (or, alternatively, of hitting a definite pin and then passing on its left side).

This new distribution will be, of course, quite different from the original distribution. It can be calculated from first principles (given a symmetrical board); and we can *test* our calculations in various ways. For example, we can let the balls roll down as usual, but list

Δq_2, the two probabilities themselves (rather than their amplitudes) are to be added and normalized: we cannot imitate the two-slit experiment. But this is not our problem at this stage.' Feyerabend also objects that Popper never attempts to explain interference patterns but instead emphasizes 'again and again the relational character of the probabilities'. But Popper was not concerned to *explain* the interference patterns: *that* is the task of quantum theory! Popper was, rather, concerned to criticize the Copenhagen interpretation of quantum theory. Popper of course does not explain interference with the pin board example. He is pointing out with it that probability is a quality of a system not ascribable to an element thereof. Hence the whole system of probabilities can alter with a change in conditions. Popper indeed contended that this is all that is involved in the reduction of a wave packet (at present often called the 'collapse of the state vector'). To make his point, he *had* to present an example of a wave-packet reduction in a classical situation. Interference is, as remarked, explained by quantum mechanics. To elucidate the specific *sort* of interference which is present in the two-slit experiment, Popper goes a step further that Feyerabend omits to mention: he calls on Landé's application of Duane's third quantum rule to explain (or rather, to show the possibility of explaining) the wave interference patterns. See my 'Critical Study: The Philosophy of Karl Popper: Part II. Consciousness and Physics', in *Philosophia* 7, 3–4, July 1978, pp. 675–716, esp. p. 695. See also Thesis 13 below. Ed.]

separately the final positions of those balls that hit the selected pin (or that hit it and pass on its left); or else, we can at once *remove* all those balls that do not satisfy the new condition. In the first case, we merely *take note* of the new 'position measurement' of the ball; in the second case, we *select* the balls which pass through some predetermined position.

In both cases we shall get tests of the calculated new distribution: the distribution of those balls which have undergone a 'position measurement' (a case analogous to a 'state preparation').

The theory of the pin board allows us, of course, to calculate from first principles the new distributions for any pin we choose; in fact, all these new distributions are implicit in calculating the original normal distribution. For this calculation assumes that the ball will hit, with such and such a probability, such and such a pin.

9. Ninth thesis. In the case of the pin board, the transition from the original distribution to one that assumes a 'position measurement' (as described here)—whether an actual one or an assumed or feigned one—is not merely analogous, *but identical with the famous 'reduction of the wave packet'* (or 'collapse of the state vector') in quantum mechanics.[66] Accordingly, the 'reduction of the wave packet' is not an effect characteristic of quantum theory: it is an effect of probability theory in general.[67]

Take our pin board example again: given not only the topography of the board but also its inclination and a few more facts, we may look at the probability distribution as a kind of descending wave front, starting to descend when the particle enters the board through its slit Δq. There will be no interference of amplitudes: if we have two slits Δq_1 and Δq_2, the two probabilities themselves (rather than their amplitudes) are to be added and normalized: we cannot imitate the two-slit experiment. But this is not our problem at this stage. What I wish to show is this: we may calculate a probability wave, descending to the bottom of the board, and forming there a normal distribution curve very much like a wave packet.

Now if we let *one* actual ball roll down, we can look at it from various points of view.

[66]Some important arguments that strengthen the ninth thesis specifically in the light of the propensity interpretation will be found in section 4 below.

[67]Cp. my *Logic of Scientific Discovery*, section 76.

(a) We may say that the experiment as a whole determines a certain probability distribution and retains it (upon repetition) irrespective of the particular pins hit by the ball.

(b) We may say that every time the ball actually hits a certain pin (or, say, passes on its left side), the *objective probability* distribution (the propensity distribution) is 'suddenly' changed, whether or not anybody takes note of the course of the ball. But this is merely a loose way of saying the following: *if we replace the specification of our experiment by another one* which specifies that the ball hits that particular pin (or passes on its left), then we have a different experiment and accordingly get a different probability distribution.

(c) We may say that the *knowledge*, or the *information*, or the *consciousness*, or the *realization*, that a position measurement has taken place, leads to the *'collapse' or 'reduction' of the original wave packet* and to its replacement by a new wave packet. But in speaking in this way, we only say the same as we said before under (b); except that we now use subjectivist language (or a subjectivist philosophy).

Obviously, if we do not *know* which pin the ball has hit, we do not *know* by which new set of experimental conditions (propensities) we could replace, in this particular case, the old one. But whether we know this or not—we *did* know from the very start that there was such and such a probability of the ball hitting such and such a pin, and thereby changing its propensity of hitting other pins, and ultimately of reaching a certain point (or column), a, at the bottom of the board. It was on this knowledge that we based our calculation of the original probability distribution (wave packet).

Let us call our original specification of the pin board experiment 'e_1' and let us call the new specification (according to which we consider or select only those balls which have hit a certain pin, x, say, as repetitions of the new experiment) 'e_2'. Then it is obvious that the two probabilities of landing at a, $p(a,e_1)$ and $p(a,e_2)$, will not in general be equal, because the two experiments described by e_1 and e_2 are not the same. But this does not mean that the new information which tells us that the conditions e_2 are realized in any way changes $p(a,e_1)$: from the very beginning we could calculate $p(a,e_1)$ for the various a's, and also $p(a,e_2)$; and we knew that

$$p(a,e_1) \neq p(a,e_2).$$

Nothing has changed if we are informed that the ball has actually

hit the pin x, except that we are now free, *if we so wish*, to apply $p(a,e_2)$ *to this case*: or in other words, we are free to look upon the case as an instance of the experiment e_2 instead of the experiment e_1. But we can, of course, continue to look upon it as an instance of the experiment e_1, and thus continue to work with $p(a,e_1)$: the probabilities (and also the probability packets, that is, the distribution for the various a's) are *relative probabilities*: they are *relative to what we are going to regard as a repetition of our experiment*; or in other words, they are relative to what experiments are, or are not, *regarded as relevant to our statistical test*.

Take another example, a very famous one, due to Einstein, and discussed by Heisenberg[68] and by myself.[69] Take a semi-transparent mirror, and assume that the probability that light will be reflected by it is $\frac{1}{2}$. Thus the probability that light will pass through will also be $\frac{1}{2}$, and we have, if the event 'passing through' or 'transmitted' is a, and the experimental arrangement b,

$$p(a,b) = \tfrac{1}{2} = p(-a,b),$$

where '$-a$' (that is, 'non-a') stands for the event 'reflection'. Now let the experiment be carried out with one single photon. Then the probability wave packet associated with this photon will split, and we shall have the *two* wave packets, $p(a,b)$ and $p(-a,b)$, for which our equation

$$p(a,b) = \tfrac{1}{2} = p(-a,b)$$

will hold. 'After a sufficient time', Heisenberg writes, 'the two parts will be separated by any distance desired . . . '. Now let us assume that we 'find', with the help of a photographic plate, that the photon (which is indivisible) was reflected. (Heisenberg says that it is 'in the reflected part of the packet', which is a misleading metaphor.) 'Then the probability', he writes,[70] 'of finding the photon in the other part of the packet *immediately becomes zero*. The experiment at the position of the reflected packet thus exerts a kind of action (reduction of the wave packet) at the distant point occupied by the

[68]*Cp.* W. Heisenberg, *The Physical Principles of Quantum Theory*, 1930, p. 39.

[69]*Cp.* my *Logic of Scientific Discovery*, the end of section 76 (pp. 235 *f.* in the English edition).

[70]*Cp.* W. Heisenberg, *op. cit.*, p. 39; the italics are mine.

transmitted packet, and one sees that this action is propagated with a velocity greater than that of light.'

Now this is the great quantum muddle with a vengeance. What has happened? We had, and still have, the relative probabilities

$$p(a,b) = \tfrac{1}{2} = p(-a,b).$$

If we take the information $-a$ (which says that the particle has been reflected), then relative to this information we get

$$p(a,-a) = 0, \quad p(-a,-a) = 1.$$

The first of these probabilities or wave packets is indeed zero. But it is quite wrong to suggest that it is a kind of changed form of the original packet $p(a,b)$ which 'immediately becomes zero'. The original packet $p(a,b)$ remains equal to $\tfrac{1}{2}$, which is to be interpreted as meaning that if we repeat our original experiment, the virtual frequency of photons being transmitted will equal $\tfrac{1}{2}$.

And $p(a,-a)$, which is zero, is quite another relative probability: it refers to an entirely different experiment which, although it begins like the first, ends according to its specification only when we find (with the help of the photographic plate) that the photon has been reflected.

No action is exerted upon the wave packet $p(a,b)$, neither an action at a distance nor any other action. For $p(a,b)$ is the propensity of the state of the photon relative to the original experimental conditions. This has not changed, and it can be tested by repeating the original experiment.

It might be thought that it is unnecessary to repeat all this. But more recently, Heisenberg has suggested that the reduction of the wave packet is somewhat similar to a quantum jump.[71] For, on the

[71][*(Added 1980) This comparison of the collapse of the state vector with the quantum jump is not very good; although, as we shall see in section 4 below, there certainly may be a real discontinuity, it is not always present. For example, there may be the discontinuity of a photon's either being reflected or not being reflected by a semi-permeable mirror. What is usually called a quantum jump is indeed characterized by two facts. One is that the jump is more or less probable, and has a probability function, but need not necessarily occur. The other fact is that we do not see the mechanism of the quantum jump. The quantum jump itself is, from the point of view of quantum theory, inexplicable. There is simply a probability or propensity to make the jump. While in the case of a semi-permeable mirror, we have something like a mechanism of reflection or of penetration. It is correct that both mechanisms in this case need not work, but one of the two may work. And

one hand, he speaks of 'the reduction of wave-packets' as 'the fact that the wave function . . . changes discontinuously', adding, 'It is well known that the reduction of wave-packets always appears in the Copenhagen theory when *the transition is completed from the possible to the actual* . . . ', that is, when 'the actual is selected from the possible, which is done by the *"observer"* . . . '. On the other hand, he speaks on the next page of the 'element of discontinuity [in] the world, which is found everywhere in atomic physics . . . [and which in] the usual interpretation of quantum theory . . . *is contained in the transition from the possible to the actual'.*[72]

Yet the reduction of the wave packet clearly has nothing to do with quantum theory: it is a trivial feature of probability theory that, whatever a may be, $p(a,a) = 1$ and (in general) $p(-a,a) = 0$.

Assume that we have tossed a penny.[73] The probability of each of its possible states equals $\frac{1}{2}$. As long as we do not look at the result of our toss, we *can* still say that the probability will be $\frac{1}{2}$. If we bend down and look, it suddenly 'changes': one probability becomes 1, and the other 0. Was there a quantum jump, owing to our looking? Was the penny influenced by our observation? Obviously not. (The penny is a 'classical' particle.) Not even the probability (or propensity) was influenced. There is no more involved here, or in *any* reduction of the wave packet, than the trivial principle: if our information contains the result of an experiment, then the probability of this result, *relative to this information* (regarded as part of the experiment's specification), will always trivially be $p(a,a) = 1$.[74]

This explains also what is valid in von Neumann's principle, mentioned in my seventh thesis above, that if we repeat a measurement at once, then the result will be the same with certainty. Indeed, it is quite trite that if we look at our penny a second time, it will still lie as before. And more generally: if we take a 'measurement' like

that leads to a discontinuity.]

[72]*Cp.* Heisenberg, 'The Development of the Interpretation of the Quantum Theory', in W. Pauli, ed.: *Niels Bohr and the Development of Physics*, 1955, pp. 12–29; see pp. 23 *f.* (the italics are mine).

[73]This example is taken from my paper 'The Propensity Interpretation of the Calculus of Probability, and the Quantum Theory', in Stephan Körner, ed.: *Observation and Interpretation in the Philosophy of Physics*, 1957, pp. 65–70 and 88*f.*

[74][*(Added 1980) See section 4 below, where this argument is revised and, I hope, improved.]

that of the arrived photon as *defining the conditions of the experiment*, then the outcome of the repetition of this experiment is certain, by virtue of the specification of the experiment together with the trite fact that $p(a,a) = 1$.

Before proceeding to my next thesis, I will just return for a moment to the pin board. Take

$$p(x,e_1) = r$$

to be the probability of a ball's hitting the pin x, in the original experiment, and assume that we see the ball passing x *without* hitting it. Then this can be interpreted exactly as Heisenberg interprets the experiment with the semi-transparent mirror:[75] we could say (it would be very misleading) that the wave packet $p(x,e_1)$ collapses, that it becomes zero with super-luminal velocity. I hope that the point need not be further elaborated.

10. My tenth thesis is that the propensity interpretation solves the problem of the relationship between particles and their statistics, and thereby that of the relationship between particles and waves.

Dirac writes:[76] 'Some time before the discovery of quantum mechanics people [Einstein, von Laue] realized that the connection between light waves and photons must be of a statistical character. What they did not clearly realize, however, was that the wave function gives information about the probability of *one* photon being in a particular place and not the probable number of photons in that place.' And he continues with an example very much like the example discussed above of the semi-transparent mirror interacting with *one* photon.[77]

Now this application of probability theory to single cases is precisely what the propensity interpretation achieves. But it does

[75]See note 70 above.

[76]*Cp.* P. A. M. Dirac, *The Principles of Quantum Mechanics*, fourth edition, 1958, p. 9.

[77]Dirac was undoubtedly right in the important issue that every single particle is accompanied by a (propensity) wave—or by waves. (The wave automatically takes all the possibilities of the surrounding situation into account.) However, he was wrong to think that only wave trains belonging to one and the same particle can interfere with each other. This dogma (whose influence upon the irrationality of the Copenhagen interpretation can hardly be overrated) has been refuted by the invention of the laser.

not achieve it by speaking about particles or photons. *Propensities are properties of neither particles nor photons nor electrons nor pennies.* Propensity statements in physics describe properties of the situation, and are testable if the situation is typical, that is, if it repeats itself (such as in the case of the emission of light). They are, therefore, also *properties of repeatable experimental arrangements*: physical and concrete, in so far as they may be statistically tested (and lead, in the pin board case, to an actual characteristic physical arrangement of balls)—and abstract in so far as *any particular experimental arrangement* may be regarded as an instance of more than one specification for 'its' repetition. (Take the tossing of a penny: it may have been thrown nine feet up. Shall we say or shall we not say that this experiment is repeated if the penny is thrown to a height of ten feet?) It is this *relativity* of the propensities that makes them sometimes look 'unreal': it is the fact that they refer both to single cases *and* to their virtual repetitions, and that any single case has so many properties that we cannot say, just by inspection, which of them are to be included among the specifications that define what should be taken as 'our' experiment, and as 'its' repetition.

But this is true not only of all propensities or probabilities (classical or quantum-mechanical); it is also true of *all physical or biological experiments,* and it is one of the reasons why experimentation is impossible without theory: what seems to be a completely irrelevant aspect of one experiment e_1, or arbitrarily variable in its repetitions, may turn out in 'another' experiment e_2 (otherwise indistinguishable) to be part of its most important specifications. Every experimentalist can give countless examples. Some so-called 'chance discoveries' have been made by getting unwanted, or unexpected, results upon repeating an experiment, and then noticing that the change in the result depended upon some factor previously conjectured to be irrelevant, and therefore not included in (nor excluded by) the specification of the experiment.

Thus the *relativity to specification* of which we have spoken is characteristic neither of quantum experiments nor even of statistical experiments: *it is a permanent feature of all experimentation.* (And a propensity relation might be regarded, and intuitively understood, as a generalization of a 'causal' relation, however we may interpret 'causality'.) For this reason it seems to me mistaken to regard statistical laws, statistical distributions, and other statistical entities,

as non-physical or unreal. Probability fields are physical, even though they depend on, or are relative to, specified experimental conditions.[78]

11. My eleventh thesis is this: even though both the particles and the probability fields are real, it is misleading (as Landé rightly insists) to speak of a 'duality' between them: the particles are important *objects* of the experimentation; the probability fields are propensity fields, and are as such in the first instance important *properties of the experimental arrangement* of the specified conditions. Of course they may in their turn become objects of study, just like other physical properties, such as the momentum of a particle. As we have seen, they may even be objects which one can kick and which can kick back. There is however no duality in any sense in which we may speak either of particle or of wave but not of both at once—exactly as there is no duality between a particle and its momentum that prevents us from speaking of both at the same time.

A simple example may illustrate that we must not take the propensity as a simple property of the particle alone but of the particle and the total situation—a situation which may occur of course without any human activity, although it also may be a situation prepared for a special study by men, especially when studying these matters.[79] One is easily tempted to look upon the probability ½ as a property of a thing, of a kind of 'particle'. *But this temptation must be resisted.* For let us assume an experimental arrangement in which the penny is not spun but tossed in such a way that it falls on a table with some slots in which it can be caught upright. Then we may distinguish between *three* possibilities: heads showing up; tails showing up; and neither showing up. Or even four possibilities: if the slots are all north-south, we may distinguish the caught pennies by the direction in which their heads face (east or west). This shows that conditions other than the structure of the penny (or the particle) may greatly contribute to the probability or propensity: the whole experimental arrangement determines the 'sample space' and the probability distribution. (We also can easily conceive of specifications according to which the experimental

[78]Cp. my *Conjectures and Refutations*, 1963, 1972, pp. 213 *f.* See also the warning in note 63 above.

[79]This example is taken from my paper 'The Propensity Interpretation of the Calculus of Probability, and the Quantum Theory', *op. cit.*, p. 89.

conditions change, perhaps even in a certain 'random' manner, while the experiment proceeds.)

Thus the propensity or probability is not (like baldness, or charge) a property of the member of the population (man, particle) but somewhat more like the popularity (and consequently, the sale statistic) of a certain brand of chocolate, depending on all kinds of conditions (advertisement, sales organization, statistical distribution in the population of preferential taste for various kinds of chocolate). And a wave-like distribution of a probability (or a probability amplitude) is, indeed, something which cannot be said to be an alternative 'picture' of the member of the population (man; bar of chocolate; particle). It would be awkward to speak of a *'duality'* (a *symmetrical* relation) between a bar of chocolate and the shape of the distribution curve of its propensity to be sold tomorrow.

[The oldest and best known example of a propensity in atomic physics is of course radioactivity, interpreted as a propensity of each radioactive nucleus to disintegrate. This propensity is extremely difficult to 'kick' and it was thought for a long time that it is in principle unkickable; that is, independent of all experimental conditions. It was only with the theory of nuclear resonance that this view has changed, and that we see that a nuclear propensity to disintegrate may, both in single atoms and in bulk, be influenced by experimental conditions. (The insertion of this paragraph is due to a reminder of Sir Hermann Bondi to whom I feel greatly indebted.)]

12. My twelfth thesis is that the mistaken idea of a duality of particle and wave is, partly, due to the hopes raised by de Broglie and Schrödinger of giving a wave theory of the *structure of particles*.

There was a span of two years between the beginning of wave mechanics and the successful analysis and interpretation of experiments as tests of Born's statistical interpretation, first presented in 1926, of the ψ-function.[80] In these years, the statistical problems seemed less important than the hope of solving the problems of atomic stability (and of quantum jumps) by a classical method. It was a very beautiful method, and an inspiring hope: the hope was

[80]*Cp.* M. Born, 'Bemerkungen zur statistischen Deutung der Quantenmechanik', in F. Bopp (ed.), *Werner Heisenberg und die Physik unserer Zeit*, 1961, pp. 103–118; see p. 104.

nothing less than one of explaining matter and its structure by field concepts. When Schrödinger and Eckart later showed the (far-reaching though not complete) equivalence of the wave theory and Heisenberg's particle theory, the two-picture interpretation was born, with its idea of a symmetry or duality between particle and wave. But in so far as there was an equivalence, it was one between *two statistical theories*—a statistical theory ('matrix mechanics') which started from the statistical behaviour of particles, and a statistical theory which started from the wave-like shape of certain probability amplitudes. We might say (being wise after the event) that Schrödinger's hope that what he had found was a wave theory of the structure of matter should not have survived[81] the successful tests of Born's statistical interpretation of the wave theory.

13. My thirteenth and last thesis is this. Both classical physics and quantum physics are indeterministic.[82] The peculiarity of quantum mechanics is the principle of the superposition of wave amplitudes—a kind of *probabilistic dependence* (called by Landé 'interdependence') that has apparently no parallel in classical probability theory. To my way of thinking, this seems to be a point in favour of saying that propensities are physical and real (though *virtual*, as stressed by Feynman). For the superposition can be kicked: coherence (the phase) can be destroyed by the experimental arrangement.

Alfred Landé has made a most interesting and it seems at least partly successful attempt to explain this peculiarity by showing this mathematically:[83] 'The question of *why do the probabilities interfere* can . . . be answered: They have no other choice if they "want" to obey a *general interdependence law* at all.' Let us assume that Landé's brilliant derivations of quantum theory from non-quantal principles of symmetry stand up to critical analysis: even then it

[81]Cp. E. Schrödinger, 'The General Theory of Relativity and Wave Mechanics', in *Scientific Papers Presented to Max Born*, 1953, pp. 65–74.

[82]Cp. my paper, 'Indeterminism in Quantum Physics and in Classical Physics', *British Journal for the Philosophy of Science* 1, 117–133 and 173–195 (1950); and *The Open Universe: An Argument for Indeterminism*, Vol. II of the *Postscript; cp.* A. Landé, 'Probability in Classical and Quantum Theory' in *Scientific Papers Presented to Max Born*, 1953, pp. 59–64; M. Born, 'Bemerkungen zur statistischen Deutung der Quantenmechanik', (see footnote 80 above).

[83]Cp. A. Landé, *New Foundations of Quantum Mechanics*, 1965, p. 82; the italics are partly mine.

seems to me that his own arguments show that these probabilities (propensities) whose amplitudes can interfere should be conjectured to be *physical and real, and not merely a mathematical device* (as he sometimes seems to suggest). Though their mathematical 'pictures' may have the shape of 'waves' only in 'configuration space', *as propensities* they are physical and real, quite independently of the question whether or not they can be represented by a wave picture, or a function with a wave shape, or, indeed, by any picture or shape at all. The wave picture may thus have only a mathematical significance; but this is not true of the laws of superposition which express a real probabilistic dependence. I therefore think that the way in which quantum mechanics differs fundamentally from classical physics—that is, in the interference of the propensity waves—shows that the propensity waves can interact and are therefore real. This is a powerful argument for the existence of propensity fields; and if it is accepted we are able to explain the most characteristic peculiarities of quantum theory.

On the other hand, it seems to me clear from the Compton-Simon photographs that *photons can be kicked and can kick back,* and are therefore (in spite of Landé's sceptical views as to their existence) *'real'* in precisely the sense which Landé himself has given to the term. [Admittedly, they are less real than, say, electrons, because photon numbers are not conserved while electron numbers are conserved at least in what may be called 'ordinary' processes (not, of course, in pair annihilation).[84] But after all, tables and chairs and human bodies are not preserved either (especially not their numbers); nor are atoms.]

As always, nothing depends on words, but talking of 'dualism of particle and wave' has created much confusion, as Landé rightly emphasizes: so much so that I wish to support his suggestion to abandon the term 'dualism'. I propose that we speak instead (as did Einstein) of the particle and its *'associated'* propensity fields (the plural indicates that the fields depend not only on the particle but also on other conditions), thus avoiding the suggestion of a symmetrical relation.

Without establishing some such terminology as this (*'association'* in place of *'dualism'*) the term 'dualism' is bound to survive, with all

[84][See *The Self and Its Brain*, 1977, p. 7. Ed.]

the misconceptions connected with it; for it does point to something important: the association that exists between particles and fields of propensities ('forces', decay propensities, propensities for pair production, and others).

Incidentally, among the misleading fashionable terms of the theory is the term 'observable'.[85] It suggests something that does not exist: all 'observables' are *calculated and inferred on theoretical grounds*, rather than observed or directly measured. Thus what is 'observable' always depends upon the theory we use. However, here again one should not quarrel about words; no more about the word 'observable' than about the word 'real'. Definitions, as usual, lead nowhere; but most of us know what we mean when we say that *there are* such things (observable things) as elephants, or electrons, or magnetic fields or (more difficult to observe) propensities, such as the propensity to attract, or to understand, or to criticize; or the propensity of an experiment to yield some specified result.

To sum up. The alleged dualism of particle and wave and the subjective interpretation of probability, with which it is closely connected, are responsible for the subjectivistic and anti-realistic interpretation of quantum theory and for such characteristic statements as Wigner's, who says that[86] 'the laws of quantum mechanics itself cannot be formulated . . . without recourse to the concept of consciousness'; a view that he attributes also to von Neumann; or Heisenberg's statement: 'The conception of objective reality . . . has thus evaporated . . . into the transparent clarity of a mathematics that represents no longer the behaviour of particles but rather our knowledge of this behaviour.'[87] Or his assertion[88] that if the observer is exorcized, and physics made objective, the ψ-function 'contains no physics at all'.

[85]*Cp*. D. Bohm and J. Bub, 'A Proposed Solution of the Measurement Problem in Quantum Mechanics by a Hidden Variable Theory', *Reviews of Modern Physics* **38**, 1969, pp. 453–69; see especially pp. 465 *f*.

[86]*Cp*. E. P. Wigner, 'The Probability of the Existence of a Self-Reproducing Unit', in M. Polanyi, *The Logic of Personal Knowledge*, 1961, pp. 231–8; see p. 232.

[87]*Cp*. W. Heisenberg, 'The Representation of Nature in Contemporary Physics', *Daedalus* **87**, 1958, pp. 95–108.

[88]*Cp*. W. Heisenberg, 'The Development of the Interpretation of the Quantum Theory', in W. Pauli (ed.), *Niels Bohr and the Development of Physics*, 1955, pp. 12–29; see p. 26.

I have often argued in favour of the evolutionary significance of consciousness, and its supreme biological role in grasping and criticizing ideas.[89] But its intrusion into the probabilistic theory of quantum mechanics seems to me based on bad philosophy and on a few very simple mistakes. These, I hope, will soon be forgotten, while the great physicists who happened to commit them will be forever remembered by their marvellous contributions to physics: contributions of a significance and depth to which no philosopher can aspire.

4. *The Collapse of the Wave Packet without Observation.*

* (Added 1980) The argument of thesis 9 is, essentially, that of my *L.Sc.D.* of 1934. It deals especially with the so-called 'reduction of the wave packet' or, what is the same, the 'collapse of the state vector', and it can be much strengthened.

As it stands, it refers to the impact of any new *information* on a probability: it shows that a probability $p(a,b)$ may be replaced, on the new information c, by $p(a,bc)$. And it shows that, if a is the final position of a ball on the pinboard, the information c that the ball has actually hit a certain pin may of course considerably alter the probability of a; it may, for example, reduce it to zero.

In this argument, the term 'information' plays a considerable part; and for this reason, one may say that it has a certain subjectivist flavour. It may seem that 'the observer' has not been completely exorcized. The main purpose of the present section is to show that, with the help of the propensity interpretation of probability (which I did not develop until shortly before 1953)[90], the exorcism is total.

However, before proceeding to show this, I wish to say something in defence of the objectivity of the argument in thesis 9. The information which plays a part in that section and which leads to the reduction of the 'wave packet' is information about *objective facts,*

[89][See *Objective Knowledge*, 1972; and *The Self and Its Brain*, 1977. Ed.]

[90]*Cp.* my 'Philosophy of Science: A Personal Report', delivered in 1953, and printed in C. A. Mace, ed.: *British Philosophy in the Mid-Century*, 1957, esp. pp. 188 *ff.*; reprinted in *Conjectures and Refutations*, 1963. *Cp.* also my 'The Propensity Interpretation of the Calculus of Probability, and the Quantum Theory', in Körner, ed.: *Observation and Interpretation, op. cit.*

facts which are quite independent of any 'observer': the *fact* whether or not, for example, a ball has hit a certain pin of a pin board. Thus the objection that, by speaking of information, I have illicitly brought in the observer, can hardly be upheld. Nevertheless, we can get rid of any reference, direct or veiled, to information, or to knowledge, or to anything that has any subjectivist flavour.

Take the example of the pin board again. The 'regular pin board' which I am using here consists in a series of lines of pins; and each pin within each line has a constant distance from every other pin within that line. The placing of the pins in each successive line is staggered, so that its pins are placed midway with respect to those on the preceding line. Each ball is of such a size that it can just pass through the constant distance between the pins.

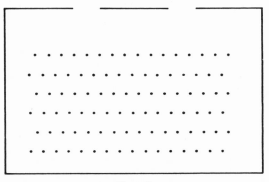

Figure 3

The regular pin board is of considerable importance because it can serve as an extremely simple model of a probabilistic world: a world in which every occurrence—that is, a ball hitting a pin—gives rise to a probabilistic choice of occurrences with probability $\frac{1}{2}$: a wave packet or a state vector representing two possible and equiprobable states, each of which has again the same character, and so on as long as we like, up to a 'final' line, in which we choose a certain pin. Hitting this pin is our occurrence a; $p(a,x)$ is the probability of hitting this chosen pin a when the situation is x.

Now the propensity interpretation regards propensities as objective physical properties of the physical situation under consideration and, ultimately, of the whole physical world. Accordingly, the objective propensity of the occurrence a—described by $p(a,x)$, a

87

wave packet or a state vector—changes whenever a ball has 'decided' to move left of some pin. It changes because x changes with every occurrence objectively.

Whether or not we are *'informed'* of the occurrence (of the value of x) is irrelevant: it is the objective propensity of a which may become zero, with or without our knowledge.

So here it is the physical world itself—the propensity—which changes, and whose change is described by the 'collapse of the state vector'. And this is just a characteristic that distinguishes a probabilistic world from a deterministic world—a world in which there are no alternatives; or to put it differently, in which all objective probabilities or propensities are either 1 or 0.

I believe that this trivial model world, the pin board, is in many respects very similar to the world which quantum mechanics depicts. There is, in the main, only one difference: in the model universe, we have no phase and no coherence of phase. Thus we have no interference of probabilities: our probabilities can add, but not their amplitudes—as was stressed in my old theses 8 and 9 above.

We can, of course, try to simulate the two-slit experiment by having balls come down the pin board through, say, the two slits A and B. But this will simply lead to the additive superposition of two Gaussian distributions rather than to the superpositions of the amplitudes of the probability waves. The reason is obvious: we cannot in our model simulate anything like phase and coherence.

One thing which our model shows very clearly is the time symmetry, that is to say, the fact that the direction of time may be reversed in our model world. If we consider any particular occurrence—any particular hitting of some pin by some ball—then we can not only ask for the distribution of the probabilities for the ball's progress, but also for the distribution of the probabilities for the ball's past course. We see at once that these two problems are exactly symmetrical: just as there is a probability of ½ for a ball B that has hit a pin c to hit either of the two nearest pins in the next line of pins, so there is a probability of ½ for the ball B to have hit, before hitting c, either of the two nearest pins in the previous line of pins. Thus working backwards in this way we obtain distributions which are exactly symmetrical to the (Gaussian) distributions which we obtain in working forward.

This, I think, beautifully illustrates the famous time-reversibility of the Schrödinger equation.

5. *Everett's 'Many-Worlds' Interpretation.*

* (Added 1981) These results may be applied to a discussion of the famous 'Many-Worlds Interpretation' of quantum mechanics, due essentially to Hugh Everett, III, and interestingly discussed by John Archibald Wheeler, Neill Graham, and Bryce S. DeWitt.[91]

In my opinion, Everett's contributions are excellent, although I do not think they accomplish what they are supposed to do—especially in the book edited by DeWitt and Graham.

Everett's three main contributions appear to me to be as follows:

(1) Everett's is a completely *objective* discussion of quantum mechanics.

(2) In Everett's approach (as opposed to the Copenhagen interpretation) there is no need, and no occasion, to distinguish between 'classical' physical systems, like the measuring apparatus, and quantum mechanical systems, like elementary particles and (smallish) systems of elementary particles (say, molecules or groups of molecules). Instead, all physical systems are regarded as quantum mechanical systems, especially the apparatus used in measurements; and, indeed, the universe.

(3) Everett shows that the collapse of the state vector—something that had, before him, been regarded as outside Schrödinger's theory—can be shown to arise *within* what he calls 'the theory of the Univeral [Schrödinger] Wave Function'.

The third of these points arises from my own interpretation of Everett's theory. (It is contradicted, for example, by DeWitt, who maintains that the 'collapse of the state vector' does 'not follow from the Schrödinger equations'.[92] However, perhaps DeWitt wishes only to say that it does not follow unless we adopt something like Everett's argument.)

[91]See Bryce S. DeWitt and Neill Graham, eds.: *The Many-Worlds Interpretation of Quantum Mechanics*, 1973, which includes Hugh Everett, III's paper: '"Relative State" Formulation of Quantum Mechanics', in *Reviews of Modern Physics* 29, 3, pp. 454–462, July 1957.

[92]DeWitt and Graham, eds., *op. cit.*, p. 159, last paragraph in the left-hand column.

On the first two of my three points I should expect everybody to agree that they are important aspects of Everett's treatment.

Several other aspects of Everett's treatment, or of the interpretation of his treatment by DeWitt, seem to me to be due to misunderstandings. I may mention a few of these.

There is, first of all, the so-called 'many-worlds interpretation of quantum mechanics', or the somewhat sensational thesis that the world splits at any quantum-mechanical interaction into many worlds. I shall try below to show that this is untenable.

There is, secondly, the claim that Everett's interpretation is a direct consequence of the mathematical formalism, without any additional assumption (except a 'postulate of complexity' as DeWitt calls it: 'The world is sufficiently complicated to be decomposable into systems and apparatuses'[93]). It is part of this claim that Everett's interpretation is free of 'metaphysical' assumptions. This claim, too, will be shown to be not tenable.

However, my criticism should not be taken to belittle Everett's achievement or his originality.

Everett's argument is simple and ingenious. He considers the quantum mechanics of more or less complex systems which are regarded as (sufficiently) isolated from the rest of our world. What holds for these systems may then be extended to the whole universe.

The physical systems X which are specially investigated by Everett are those which contain a subsystem A that consists of an apparatus (a computer) that can measure some of the so-called 'observables' of a subsystem B of X, and record the measurement in a physical subsystem of A, called, metaphorically, the 'memory' of A.

The subsystem B whose evolution is recorded by A may be (for simplicity) taken to be our regular pin board; and A may be taken to measure and to record the probability of a ball to proceed on one of its possible paths down the board. These probabilities will be represented by a 'wave packet' or 'state vector'; and a 'measurement' (of the position of the ball) may or may not be the occurrence[94] of the ball hitting a certain pin. If it is, then the 'measurement' will lead

[93]*Ibid.*, p. 160.
[94]*Cp.* my discussion of 'occurrence' and 'event' in *L.Sc.D.*, section 23, pp. 88–91.

to the 'collapse' of the previous state vector, in the senses described above in thesis 9 and in section 4.

In so far as the physical system A records the collapse of the state vector, it shows the physical objectivity of this collapse: the collapse is shown by Everett to be part of the objective quantum-mechanical theory.

So far so good.

But Everett has put an interpretation on his results[95] which I regard as definitely metaphysical. The motivation of this interpretation is very simple. The quantum mechanical theory, say in the form of Schrödinger's equation, does not predict (like every probabilistic theory) which of the possible paths open to it the system will choose. He says—perhaps believing that only in this way will the objectivity of the theory be preserved—that 'with each succeeding observation' the state of the system 'branches into a number of different states. Each branch represents' one of the possible 'different outcome[s] of the measurement and the *corresponding* eigenstate for the object-system [B's] state. All branches exist simultaneously . . . after any given sequence of observations.'

This view of Everett's can be illustrated by our pin board model, enriched by an observing and recording computer A. It amounts to the assertion that (as the state vector determining the next *two* possible states after any 'measured' state *objectively* determines *two* states) there will be a split of our system X (pin board B + computer A, or $A + B$) into two systems: $X' = A' + B'$ and $X'' = A'' + B''$, each of them realizing one of the predicted objective possibilities. Of course, in X', A' will record a state vector that represents the state (future possibilities) of B' while in X'', A'' will record the state vector that represents the state (future possibilities) of B''.

In other words, at every occurrence (a ball hitting a pin) our system X (i.e., board + computer) duplicates. (In a more complicated system B with more than two possibilities opened up by an interaction, the system X will split into n systems, where n is the number of possibilities opened up by the occurrence in B.)

Each of the many worlds (systems) following a split (an occurrence) represents one of the possible worlds—one of the worlds

[95]In his paper of 1957, cited in footnote 91 above.

which remain possible after the occurrence (the measurement, say) that led to the split.

Clearly, since each of these worlds is *identical* with one of the worlds that actually could have arisen according to common sense, there will be no interaction between them: no inhabitant of one of them will notice anything about the other worlds, or notice anything that should not happen in a commonsense world.

Therefore, Everett's interpretation is not made impossible by the fact that we do not experience the continual splitting of our world (which means that we ourselves split continually). It is, as he points out, no more contradicted by our failure to experience the splitting than the theory that the earth rotates is contradicted by our failure to experience its movement.

Nevertheless, I think that Everett's many-worlds interpretation cannot possibly be upheld; not because of its conflict with immediate experience, or with common sense, but because of conflicts with the physical theory which he thinks he follows faithfully.

There are many such conflicts. Some can perhaps be resolved by something very much like special pleading. But some are, in my opinion, absolutely fatal to his theory.

Of those conflicts which can be resolved by special pleading are those which are connected with the conservation laws—the conservation of energy (energy-mass), of momentum, of angular momentum, etc. These laws would be, quite obviously, drastically violated by every split.

The way out would be to say that these laws hold *within* every one of the branching systems, but not of the systems of branching systems.

This surely can be said, but there is a strong objection to this way out: the only reason for introducing the splits, and the many worlds, is that a *part of the physical theory*—the one which refers to the possibilities which a system can adopt—can be taken literally as a description of the actual total physical reality rather than of alternative possibilities between which the system can choose (as in the case of the pin board). But then, the way out mentioned says that the conservation laws must not be taken literally—not even those of them which are implied by the same Schrödinger equation which is the basis of Everett's theory.

I do not think that this way out is at all acceptable.

(Incidentally, we see very clearly, even at this stage, that the many worlds interpretation does not, as claimed, follow directly from the physical formalism.)

But there is an even stronger criticism.

The Schrödinger equation is symmetrical with regard to a reversal of the direction of time—just as our regular pin board, as has been explained at the end of section 4.

But this means that if we accept Everett's theory of splits, that is to say, the reality of every possibility with respect to the future, we have also to accept the opposite—the constantly occurring *fusion of an infinity*[96] *of real worlds* at the time of every occurrence. This is inevitable on Everett's theory, for it must preserve one of the most important symmetries of the formal theory it is supposed to interpret.

Let us denote the moment of time at which we make a state preparation or an observation t_1. The ordinary use of Schrödinger's equation is to ask what will happen at t_2 (where $t_1 < t_2$). According to Everett's interpretation, if there are two possibilities, each let us say to be predicted with the probability ½, then the world will split into two, and each probability will be realized in one of the worlds.

But we can also use Schrödinger's equation to ask out of what states, at t_0, the state at t_1 may possibly have arisen. Again the answer may be 'out of two states, with the probability of ½ out of each'. I think it is clear that Everett's interpretation would demand, in this case, for reasons of symmetry, that the state at t_1 arose from the fusion of the states at t_0. Since the formalism is symmetrical, the interpretation has to be symmetrical also.

Everett and his commentators do not say a word about this; and it is, indeed, most difficult to say anything to defend this view of the world—especially once it has been shown that the many-worlds interpretation is not a consequence (and thus not an integral part) of the quantum mechanical formalism, but a *more or less* (but admittedly not entirely) arbitrary interpretation of the formalism, and certainly a metaphysical interpretation (as is also, I freely admit, the

[96]Incidentally, it is easily seen that the number of the 'many' worlds on Everett's theory is infinite, even if the object system B is merely a hydrogen atom, and that the infinity is even of the power of the continuum.

commonsense view of the pin board, or my propensity interpretation).

But there are further difficulties raised by the time reversibility situation. One is this. The Schrödinger wave is, formally, of a deterministic character; it becomes probabilistic only through our interpretation (the Born interpretation, which corresponds to the commonsense interpretation of the pin board world). One point Everett mentions in support of his many-worlds interpretation is that it preserves the deterministic character of the Schrödinger wave interpretation: if all the possibilities that the wave opens up are realized, then indeed reality is predetermined by any state described by a solution of the wave equation.

This can be illustrated vividly by any state preparation experiment, say, by a beam of particles selected by a narrow slit.

Each particle of the scattering beam can be taken as an analogue of one of the splitting Everett worlds; and the whole beam as an analogue of the Everett reality—the many worlds which, one sees here, are not only many, but are scattering in a random manner relative to each other.

Now let us invert the direction of time. Then we see that the many worlds of the past are a random scatter,[97] yet so arranged that when they fuse they become correlated, *even though there was no interaction between them before their fusion.*

I may just add that if we consider the fusion of memories, the situation becomes clearly absurd, and most difficult to describe.

It is clear that the realistic tendency behind this theory can be fulfilled in a less absurd fashion. In fact, it is fulfilled by the propensity interpretation which, admittedly, is no more deducible from the formalism than is Everett's interpretation. But in regarding propensities as physically real (but, incidentally, invested with a

[97]We get a similar absurdity if we apply the Schrödinger equation to a gram of some radioactive material and then turn the arrow of time round so that the material integrates instead of disintegrating. The situation is possible but improbable in the extreme. See my notes in *Nature* on the Arrow of Time (*Nature* **177** (1956), No. 4507, p. 538; *Nature* **178** (1956), No. 4529, p. 382; *Nature* **179** (1957), No. 4573, p. 1297; *British Journal for the Philosophy of Science* **8** (1957), No. 30, pp. 151–155; *Nature* **181** (1958), No. 4606, pp. 402–403; *Nature* **207** (1965), No. 4994, pp. 233–234; *Nature* **213** (1967), No. 5073, p. 320; and *Nature* **214** (1967), No. 5085, p. 322.)

time direction) we can say that Schrödinger's equation describes a physical reality; and the collapse of the state vector can be 'justified' within the theory by Everett's ingenious argument.[98]

[98][Popper's propensity interpretation is presented in *Realism and the Aim of Science* (Vol. I of the *Postscript*), and throughout the present volume, as well as in *The Open Universe: An Argument for Indeterminism* (Vol. II of the *Postscript*). It was first presented in 1953, in the course of lectures published in C. A. Mace, ed.: *British Philosophy in the Mid-Century*, 1957, pp. 188 *ff.* (reprinted in *Conjectures and Refutations*, 1963, Chapter 1). See also H. D. Lewis, ed.: *Contemporary British Philosophy*, 1956, p. 388 (reprinted in *Conjectures and Refutations*, Chapter 3). Ed.]

QUANTUM THEORY
AND THE
SCHISM IN PHYSICS

UNDERSTANDING QUANTUM THEORY AND ITS INTERPRETATIONS

CRITICS of the Bohr-Heisenberg interpretation of quantum theory are, as a rule, dismissed as reactionaries who sigh for the good old days—the days of classical physics; the days when theories were not 'abstract' but could easily be visualized with the help of mechanical models; and the days of determinism. Hence I suppose I should repeat that I ardently admire new ideas, especially in physics, including ideas which have not been successful. Also, I am convinced that the 'abstract' character of a new theory is nothing against it, and that what strikes one generation of physicists as 'abstract' may no longer appear so to the next. (Many examples could be given to illustrate this.) And, as I must emphasize again, I am an indeterminist.[1]

A large part of *The Logic of Scientific Discovery* (*L.Sc.D.*) was devoted to the problems of quantum theory. But quantum theory is also an area in which my ideas have developed and changed somewhat; and I intend to present my current thinking on these issues in the pages that follow.[2] Before going into any details, I wish

[1][See *The Open Universe: An Argument for Indeterminism*, Volume II of the *Postscript to the Logic of Scientific Discovery*. Ed.]

[2][Since writing this volume of the *Postscript* in the mid 'fifties, Popper has published a number of essays developing his ideas on quantum physics. These include: 'The Propensity Interpretation of the Calculus of Probability, and the Quantum Theory', in *Observation and Interpretation*: Proceedings of the Ninth Symposium of the Colston Research Society, Bristol, ed. S. Körner, 1957, pp. 65–70 and 88–9; 'The Propensity Interpretation of Probability', in *The British Journal for the Philosophy of Science* 10, 1959, pp. 25–42; 'Philosophy and Physics', *Atti del XII Congresso Internazionale di Filosofia*, Venice, 1958, 2, 1960, pp.

briefly to state five points which indicate how my present ideas compare to my earlier ideas about quantum theory.

(1) One argument concerning quantum theory upon which I laid great stress in the text of *L.Sc.D.*—the imaginary experiment described in section 77—is invalid, and I wish to withdraw it.[3]

(2) I believe nevertheless that my early *criticism* of Heisenberg's and Bohr's interpretation of the quantum theory—now usually called the Copenhagen interpretation—is valid. It stands on its own feet. Moreover, my own invalid experiment can be replaced by another one which is valid—the famous imaginary experiment of Einstein, Podolsky, and Rosen.

(3) In addition to my old criticism of the 'Copenhagen interpretation', I uphold my *positive thesis that the problem of interpreting quantum theory is bound up with that of interpreting probability theory.* Thus quantum theory would have to be interpreted as a statistical theory about aggregates (or 'collectives') if probability theory had to be so interpreted.

(4) But if probability theory can be interpreted as a theory of *propensities*—in the sense of Volumes I and II of this *Postscript*—then quantum theory can also be so interpreted. Moreover, if we do

367–74; 'Quantum Mechanics Without "the Observer" ', in *Quantum Theory and Reality*, ed. Mario Bunge, 1967, pp. 7–44, reprinted as the Introduction to the present volume; 'Birkhoff and von Neumann's Interpretation of Quantum Mechanics', *Nature* **219**, 17 August 1968, pp. 682–5; 'Quantum Theory, Quantum Logic, and the Calculus of Probability', in *Akten des XIV. Internationalen Kongresses für Philosophie* **III**, Vienna, 1969, pp. 307–313; 'A Realist View of Logic, Physics, and History', in Wolfgang Yourgrau and Allen D. Breck, eds.: *Physics, Logic and History*, 1970, pp. 1–30, 35–7; 'Particle Annihilation and the Argument of Einstein, Podolsky, and Rosen', in Wolfgang Yourgrau and Alwyn van der Merwe, eds.: *Perspectives in Quantum Theory: Essays in Honor of Alfred Landé*, 1971, pp. 182–98; and 'A Realist View of Logic, Physics, and History', Chapter 8 of *Objective Knowledge*, 1972. See also Popper's intellectual autobiography and replies to his critics in P. A. Schilpp, ed.: *The Philosophy of Karl Popper*, 1974, pp. 71–77, 101–105, 108–110, 121–133, 1053–1058, 1125–1144. Popper's intellectual autobiography has been republished as *Unended Quest*, 1976, and references will ordinarily be to the latter edition. Two of the articles just cited—'Birkhoff and von Neumann's Interpretation of Quantum Mechanics', 1968, and 'Particle Annihilation and the Argument of Einstein, Podolsky, and Rosen', 1971—contain some points and arguments which Popper no longer upholds. See the 'Preface 1982' to the present volume, and also J. S. Bell: 'Quantum Mechanical Ideas', *Science* **177**, September 8, 1972. Ed.]

[3][See the discussion of Popper's imaginary experiment, and of its relationship to the Einstein-Podolsky-Rosen argument, in the 'Preface 1982' to the present volume. Ed.]

interpret quantum theory as a theory of physical propensities, then we can solve all those difficulties which have given rise to the Copenhagen interpretation.

(5) When, in the early 'thirties, I first attacked what is now called the Copenhagen interpretation, I was concerned by the anti-rational tendencies inherent in it. At that time I suggested that the subjective interpretation of probability was responsible for this: indeed the subjective interpretation induces its addicts to take to that even more dangerous (because even more narcotic) intellectual drug, the *mixture* of the subjective and objective interpretations of probability. I have learned a great deal since, and I have changed my mind on a fundamental point (see Volume I, Part 2), by giving up the frequency for the propensity interpretation. (Both of these are *objective* interpretations.) But I am more firmly convinced than ever that the lethal mixture of subjective and objective interpretations of probability has created all those irrationalist symptoms, such as the dream of the quantum theoretical interference of the subject with the object of knowledge.

What I did not see twenty years ago was the close connection between this mix-up and the determinist interpretation of classical physics.[4] In the last analysis it may well turn out that the unconscious determinist prejudice is chiefly to be blamed for the subjective theory of probability and its consequence—the invasion of mysticism into physics.

1. *A Schism in Physics.*

Since the autumn of 1934, when I first published *The Logic of Scientific Discovery* (*L.Sc.D.*), there have been many discussions of the problem of *interpreting the formalism of quantum theory*. Yet I do not think that much progress has been achieved, apart from a fundamentally important paper by Einstein, Podolsky, and Rosen, which will be examined in section 17 below.

One remarkable aspect of these discussions was the development of a split in physics. Something emerged which may be fairly described as a quantum orthodoxy: a kind of party, or school, or

[4][See Popper's discussion of determinism in *The Open Universe: An Argument for Indeterminism*, Vol. II of the *Postscript*. Ed.]

group, led by Niels Bohr, with the very active support of Heisenberg and Pauli; less active sympathizers were Max Born and P. Jordan and perhaps even Dirac. In other words, all the greatest names in atomic theory belonged to it, except two great men who strongly and consistently dissented: Albert Einstein and Erwin Schrödinger. For a time, after proposing his unorthodox theory of pilot waves, Louis de Broglie seemed to have joined the orthodoxy; but he later dissented again, encouraged by David Bohm's rediscovery of these pilot waves. And Alfred Landé, who for many years subscribed to Bohr's position, later also dissented for quite different reasons; in fact, merely in consequence of a determined effort to *understand*.

Unlike the orthodoxy, the dissenters are far from united. No two of them agree (except perhaps de Broglie and Bohm).

There is even a third group of physicists—perhaps the majority. It consists of those who have turned away from these discussions because they regard them, rightly, as philosophical, and because they believe, wrongly, that philosophical discussions are unimportant for physics. To this group belong many younger physicists who have grown up in a period of over-specialization, and in the newly developing tradition of the cult of narrowness, and the contempt for the non-specialist older generation (whose members seem to them quite antediluvian): a tradition which may easily lead to the end of science and its replacement by technology.

What is the significance of this dissension? There are two closely connected issues. ' . . . All physicists agree', writes Pauli,[1] 'that the present quantum theory, which is insufficient to explain the atomistic nature of electricity and . . . the mass values of the "elementary" particles . . . can only have a limited range of application.' In other words, all physicists agree that a new theory is needed, in spite of the marvellous success of the quantum theory. Some admit it grudgingly, while some dissenters point out that the present theory is, in fact, refuted at a number of different points, though none of the refutations is unambiguous. (Refutations are rarely quite unambiguous: we may remember the perihelion motion of Mercury.) Some dissenters feel that those who belong to the orthodoxy are inclined to belittle the significance of these points, or to cover them up by re-interpretation. This is one of the issues.

[1] 'Editorial', *Dialectica* **2**, 1948, p. 311; *cp.* also *Les Prix Nobel en 1946*, p. 146.

2. *The Significance of Interpretations.*

The second issue is perhaps more important. Since all physicists, as Pauli says, hope for a reform of the theory, there is the problem of how to search for it. There are, in the main, two methods. One is to look at the mathematical formalism and try to alter it, and if possible to generalize it. The other is to try to interpret the formalism, to understand it physically, in the hope of gaining a better understanding of its physical difficulties and shortcomings.

These two methods (if they may be so called) are not necessarily in conflict. Yet the main issue at stake is doubtless the issue between them. For the orthodoxy has adopted a view which is compatible only with the first method: tinkering with the formalism. As opposed to this, every one of the dissenters has proposed new physical interpretations in an attempt to *understand* the physical situation.

If we wish to look more closely at this last and, I believe, most important issue—the problem of *understanding* our theories—then we may discern that the orthodox party represents, in its attitude towards quantum theory, a philosophical theory of the nature of science which implies the futility of the dissenters' attempt to understand. It is the view that there is nothing there to be understood: that we can do no more than *master the mathematical formalism,* and learn *how to apply it.* This view, which Bohr and other members of the orthodoxy have laboured to establish, is practically taken for granted nowadays by the third group, to which most of the younger and tougher physicists belong. I have called this view '*instrumentalism*', or, more fully, '*the instrumentalist interpretation of scientific theories*'.[1] It is the view that all, or some, scientific theories are *nothing but* mathematical formalisms which have useful applications, in particular for the prediction of the results of experiments. The emphasis is here on 'nothing but'; for everybody, even the most ardent realist, admits of course that theories are *also* formalisms, and that they are marvellous instruments for predictions and for other applications.

Thus the instrumentalist view is widely taken for granted today; but its victory was prepared by Bohr's interpretation of the quantum theory.

[1][See Popper: *Realism and the Aim of Science,* Vol. I of this *Postscript,* sections 10–16; also *Conjectures and Refutations,* Chapter 3. Ed.]

Not everybody is satisfied with this situation. It is especially unsatisfactory to those who see in physics not mainly an instrument for predictions, or other practical applications, but rather an instrument (if instrument it must be) for understanding the world we live in—for *explaining* this world. This became very clear in, for example, the long discussion between Bohr and Einstein, lasting from 1927 to 1949,[2] and continued by Einstein until 1953.[3] Realists such as Einstein and Schrödinger feel that theories are not only instruments but are also attempts to describe a physical reality. 'I (believe) in a world of something objectively existing which I try to catch in a wildly speculative way', Einstein once wrote to Born.[4] Of course Einstein believed that his 'wildly speculative' theories should be severely tested, and for this reason, they *had* to be instruments for prediction. But what he was after (though he knew he would hardly catch it) was the real world, of which he hoped to give a true description.

Nothing is easier than to denounce statements of this kind as metaphysical or philosophical or meaningless. 'What we want are facts and figures', we hear from all sides, 'not philosophical discussions.' Some hard-boiled physicists of the present generation seem to have no interest in such discussions; or else they leave them to Niels Bohr. What they are interested in are (i) *a formalism* and (ii) *its application*; and they are interested in nothing else.

But this instrumentalism, this fashionable attitude of being tough and not standing any nonsense—especially any philosophical nonsense—is itself an old philosophical theory, however modern it may seem to us. For a long time the Church used the instrumentalist view of science as a weapon against a rising science. 'You may be able to forge instruments, gadgets, but you cannot by the light of your own reason discover any of the hidden secrets of this world': this, in brief, was the argument with which Cardinal Bellarmino opposed Galileo's teaching of the Copernican system, and with which Bishop Berkeley opposed Newton.

[2] *Cf.* Niels Bohr, 'Discussion with Einstein on Epistemological Problems in Atomic Physics', in *Albert Einstein: Philosopher-Scientist,* edited by P. A. Schilpp, 1949.

[3] *Cf.* A. Einstein, 'Elementare Ueberlegungen zur Interpretation der Grundlagen der Quantenmechanik', in *Scientific Papers Presented to Max Born,* 1953. This must have been one of Einstein's last papers, if not the last.

[4] In a letter dated 7 November 1944, quoted by M. Born in his *Natural Philosophy of Cause and Chance,* 1949, p. 122.

I have told a little of this story in another place,[5] where I also tried to show that the instrumentalist philosophy is a creed liable to be used in a defensive mood—in an attempt to *escape refutations*. For an instrument raises no claim to truth, and so cannot be falsified in the sense in which a theory which does raise such claims can be falsified.

Thus instrumentalism only revives a philosophy of considerable antiquity. But modern instrumentalists are, of course, unaware that they are philosophizing. Accordingly, they are unaware of even the possibility that their fashionable philosophy may in fact be uncritical, irrational, and objectionable—as I am convinced it is.

I have no intention, however, of discussing quantum theory, or the schism in physics, in the light of the problem of realism *vs.* instrumentalism. First, I have dealt with this aspect elsewhere. Secondly, I wish to treat the problem here under a different aspect—that of probability theory. These two aspects are, of course, related. A concern with physical interpretation, as opposed to tinkering with formulae, leads to an attack on the problem of an objective, and even a physical, interpretation of probability.

In order not to be misunderstood, I must make clear what I do *not* mean by physical understanding, and by a physical interpretation. I certainly do not mean by it the building of models, and least of all the building of models in terms of a superseded theory (such as mechanical models). Nor do I mean working with pictures or metaphors which may give us a feeling of familiarity or ease. We should remember that *forces* were not easily intuited by Newton and his contemporaries, and that for a long time physicists found Faraday's and Maxwell's *fields* of forces very difficult to understand. It is not intuitive ease I am after, but rather a point of view which is sufficiently definite to clear up some difficulties, and to be *criticized in rational terms*. (Bohr's complementarity cannot be so criticized, I fear; it can only be accepted or *denounced*—perhaps as being *ad hoc*, or as being irrational, or as being hopelessly vague.[6])

[5]*Cp.* my paper 'Three Views Concerning Human Knowledge', in *Contemporary British Philosophy* 3, edited by H. D. Lewis, 1956. See also my 'A Note on Berkeley as a Precursor of Mach', *Brit. Journ. Phil. Science* 4, 1953, pp. 26*ff*. [Both essays are reprinted in *Conjectures and Refutations*, 1963. Ed.]

[6]Einstein spoke of it as 'Bohr's principle of complementarity, the sharp formulation of which . . . I have been unable to achieve despite much effort which I have expended on it.' (P. A. Schilpp, *op. cit.*, p. 674.) The theory that Einstein was

There is not much point, however, in trying to explain in abstract terms what I mean by a physical interpretation of quantum theory; for I have offered one in *L.Sc.D.* (Chapter IX)—one which could be and has been criticized—and I am now going to offer a different one in terms of the propensity theory of probability. It is an interpretation which is not *ad hoc*, for it applies to coin tossing and to classical statistics as well as to quantum theory. And it too can be criticized. I hope it will be.

3. *Subjective Probabilities, Statistical Probabilities, and Determinism.*

When writing *The Logic of Scientific Discovery*, I was unaware of the possibility of a propensity interpretation: I believed, not without good reasons, that the only acceptable interpretation of probability in physics was the *'frequency interpretation'* or *'statistical interpretation'*, of the type given by Venn or von Mises; and throughout this *Postscript* I have continued using the terms 'statistics' and 'statistical interpretations' in this sense. From the point of view of the statistical interpretation, *singular* probability statements must be said to be *singular in form only* (or 'formally singular': cf. *L.Sc.D.*, section 71); they are in fact to be interpreted with the help of *statistical statements* about populations or collectives of a certain kind.[1]

I always realized, of course, that many physicists (Boltzmann, Einstein, and especially the quantum theorists), even while speaking of 'statistics', were regularly using *singular* probability statements; but I thought that this fact confirmed my analysis, in so far as these singular statements were to be interpreted as merely 'formally singular'.

I now think that I was mistaken, and that, in using singular probability statements, at least some of these physicists were groping, more or less consciously, for a kind of propensity interpretation. But those of them who were, intuitively, determinists

not clever enough to understand this principle, or too old fashioned ('antediluvian' and 'regressive' or 'retrogressive' are the fashionable and progressive terms applied to him) seems to me a little *ad hoc*.

[1][See Popper's discussion of probability in Volume I, Part 2 of this *Postscript*. Ed.]

were prevented by their determinist belief from reaching clarity with respect to the propensity interpretation; for, as I have argued,[2] propensities can be accepted as physical realities (analogous to forces) only when determinism has been given up.

I was also mistaken in another way: I did not see with sufficient clarity that Max Born, when introducing his 'statistical interpretation' of quantum theory, used singular probability statements which, although they could be interpreted as merely 'formally singular', *he seriously intended as singular*—viz., as singular *subjective* probability statements, as statements about our ignorance. Since I was never a determinist myself (although I always knew that theories of a character which I now call *'prima facie* determinist' can be highly informative and highly testable), I did not yet see how severely a determinist—and even an ex-determinist—was tempted to interpret all probability statements *subjectively* (or logically), that is to say as *statements of our partial ignorance.* I did not realize that he might nevertheless call them 'statistical' because he believed that he could derive (or 'almost deduce') his statistical conclusions from these singular probability statements in the manner of Bernoulli.

I knew that such derivations were palpably invalid, as explained in *L.Sc.D.*, section 62. At first, I simply re-interpreted these cases in the statistical sense without trying hard enough to understand the motives behind the subjective probability interpretation. I was content to point out, repeatedly, that no objective statistical statements could be derived from a subjective theory of ignorance (that of Jeffreys, for example, or Keynes).

Today I can see why so many determinists, and even ex-determinists who believe in the deterministic character of classical physics, seriously believe in a subjectivist interpretation of probability: it is, in a way, the *only reasonable possibility* which they can accept; for objective physical probabilities are incompatible with determinism; and if classical physics is deterministic, it must be incompatible with an objective interpretation of classical statistical mechanics.

(I may neglect here the other possibility, alluded to by Landé in the passage quoted in *The Open Universe: An Argument for*

[2][See *The Open Universe: An Argument for Indeterminism,* Volume II of this *Postscript,* sections 27–30. Ed.]

Indeterminism, section 29: using the objective hypothesis of molecular chaos. For when it is realized that this assumes a pre-established harmony—which keeps the probabilities of heads and tails constant—or alternatively a kind of teleological principle—explaining the effect of mixing processes—nobody is likely to entertain it.)

The subjective interpretation of probability seems to work as long as one fails to see the invalidity of the 'bridge' which leads from it to objective frequencies. This bridge is, of course, the law of large numbers, or the strong law of large numbers. But these laws cannot be derived in the subjective theory because their derivation is based upon an assumption of independence which (as Keynes saw) contradicts the subjective interpretation.[3]

And indeed, it is clearly absurd to believe that pennies fall or molecules collide in a random fashion *because we do not know* the initial conditions, and that they would do otherwise if some demon were to give their secret away to us: it is not only impossible, it is absurd to explain objective statistical frequencies by subjective ignorance.

The absurdity makes itself felt at times; when it does, then the physicist will, as a rule, adopt an objective view. Accordingly, we observe a characteristic oscillation, a wavering on the part of many physicists, between the subjectivist and the objectivist interpretations. As a consequence of this wavering, they lose their sensitivity to these issues (which are felt to be philosophical anyway); and bold assertions belonging to the objective theory may be found side by side—in fact, on the same page—with equally bold assertions of subjectivism.

This systematically wavering attitude—that is to say, the systematic attempt to have it both ways—is to be found in the Copenhagen interpretation. But before turning to quantum theory, I shall first briefly discuss the situation in classical statistical mechanics.

4. *The Objectivity of Statistical Mechanics.*

Statistical mechanics provide many striking illustrations of how probability statements which ought to be interpreted objectively

[3]See the discussions in *Realism and the Aim of Science*, Vol. I of this *Postscript*, Part 2, sections 3, 4, and 7.

(either purely statistically, or in the sense of the propensity interpretation), are often misinterpreted in the sense of a subjective theory of our ignorance.

Consider a big evacuated vacuum flask into which we put a little medicine bottle filled with air. We then uncork the bottle. We all know what will happen—objectively—in every case: the air will escape from the bottle and will quickly become distributed evenly throughout the flask.

We also know that even if we wait for a very long time, we shall never find an instance of the reverse process: the air will not spontaneously return into the bottle. We are faced with an irreversible process.

This simple fact has given rise to interminable discussions mainly because it is *not,* of course, physically impossible for the air to return. If a molecule moves in one direction, it must be physically possible for it to move in the opposite direction also; but if the direction of the movements of all the molecules are reversed, then they must withdraw into the bottle.

Since the physical possibility of this event cannot be doubted, we explain the experimental fact that the process is irreversible by the extreme *improbability* of a spontaneous reversion into the bottle. And since the fact to be explained—the irreversibility of the process—is an objective experimental fact, the probabilities and improbabilities in question must be objective also.

An objective probabilistic explanation of irreversible processes of the kind described is clearly outlined by Planck. He indicated that 'the overwhelming majority of cases', or more precisely, of all possible *initial states* (where each 'initial state' comprises all the initial conditions for all the molecules) was of such a kind that 'the events resulting from them agree completely in their mean value (though not in all details) while those cases which lead to noticeable deviations [from the mentioned mean values] occur only in . . . vanishing numbers; for they occur only if certain very special . . . conditions obtain—conditions which relate the position and velocities of the molecules in very special ways . . .'.[1] That is to say, the very special initial conditions which would produce noticeable deviations from normal behaviour have a vanishing probability.

[1]See M. Planck, *Theorie der Wärmestrahlung*, 5th ed., 1923, §116, pp. 114 f.

The method by which we can show that the measure of the set of initial states which produce normal states equals one, and that the measure of initial states which produce non-normal states equals zero, is fundamentally identical with the one envisaged by Bernoulli: just as we can deduce from the assumed equiprobability (equal measure of possibility) of heads (0) and tails (1) and the independence of successive tosses, the stability of the averages of almost all sufficiently long sequences (the averages will be very nearly equal to ½), so we can deduce, from simple assumptions concerning equiprobability of possible states, that almost all states will 'agree completely in their mean values', as Planck puts it, i.e., that they have stable averages with respect to distribution of the positions and velocities of the molecules.

We may attach the label 'normal' or 'approximately in equilibrium' to any state for which the positions and velocities (momenta) of the molecules are so distributed that there are no macroscopic fluctuations; or more precisely, that their averages over any small but macroscopic sub-volume do not appreciably differ from their average over the whole volume. It can then be shown that the measure of the set of normal states approaches one; and that of the non-normal states, zero. Since, moreover, the measure does not vary if, owing to the movements of the particles, the various states belonging to it change into other states, we find that almost all states must be such that, as the result of these movements, they change again into normal states (so that the time averages of almost all states are again normal[2]). Fluctuations will of course occur, but they will be too rare and too brief to affect the time averages (except for a set of measure zero).

In this way we find that there is a probability indistinguishable from zero of finding a small bottle from which the air will not escape if it is uncorked in a large vacuum; and, *vice versa*, that the probability is zero of finding a case in which the molecules will spontaneously withdraw into the small bottle.

All these considerations operate, of course, with a strictly objective (frequency or propensity) theory of probability: they are as objective as the irreversible process which is to be explained, and hold irrespective of our own state of knowledge or ignorance with

[2]*Cf.* E. Hopf, *Ergodentheorie*, in *Ergebnisse der Mathematik*, Springer, 1936.

respect to the actual detailed initial conditions of the molecules involved. To put it still more explicitly, they hold whether we are ignorant, as in fact we always are, of the initial conditions of all molecules, or whether we are supplied by a demon with full information about all these initial conditions, and thus about the precise state of the gas.

5. *The Subjectivist Interpretation of Statistical Mechanics.*

Such, however, is not the usual view of the matter. As indicated in section 3, many physicists—among them some of the most famous—try to explain these matters in terms of a subjective theory of probability. They assert that irreversibility is a result of our ignorance of the details of the state of the gas.[1]

They interpret the tendency or propensity of the system to increase its *disorder* (or '*entropy*')—a tendency which the statistical theory tries to explain on Bernoullian lines—as a tendency of our *ignorance* to increase. That is to say, instead of interpreting the 'entropy' of a system as a measure of its objective state of disorder or randomness, they interpret it as a measure of our own subjective state of ignorance of the system. This interpretation leads to the absurd result that the molecules escape from our bottle because we do not know all about them, and because our ignorance is bound to increase unless our knowledge was perfect to begin with.

I believe that this is palpably absurd, and that hot air will continue to escape even if there is nobody in the quad to provide the necessary nescience.

These absurd views are nonetheless implicit in the theories of some famous physicists who have made very important contributions to quantum theory. I shall first quote from a critical analysis of the status of probability taken from a paper on 'Probability and Physics', by W. Pauli.[2] He writes: 'The first application of probability in physics—and one which is fundamental for our understanding of natural laws—was the general statistical theory of heat, founded by Boltzmann and Gibbs. As is well known, it led by

[1][See also section 36 of Popper's intellectual autobiography, in P. A. Schilpp, ed.: *The Philosophy of Karl Popper*, 1974, and in *Unended Quest*, 1976. Ed.]

[2]*Cp.* W. Pauli, 'Wahrscheinlichkeit und Physik', *Dialectica* **8**, 1954, p. 112–24. The quotation is from pp. 114 *f*.

necessity to an interpretation of the entropy of a system as a characteristic of its state which (in contradistinction to energy) depends upon our *knowledge* of the system.'

Leaving no doubt about his subjectivism, Pauli continues: 'If this knowledge of ours is maximal—that is to say, the most precise knowledge which is at all compatible with the laws of nature . . . then the entropy is always zero. Thermodynamic concepts, however, are applicable to a system only if our knowledge of its initial state is inexact'

Pauli clearly does not realize that this subjectivist interpretation is the offspring of determinism. On the contrary, having himself become used to the subjectivist interpretation in the field of quantum theory, which is non-deterministic, he finds it necessary to mention that the subjectivist interpretation does not contradict determinism:

'The application of the concept of probability in physics just described,' Pauli continues, 'although a fundamental one, was logically compatible with a deterministic form of the laws of nature.' The remark on compatibility is correct, but the words *'although a fundamental one'* betray a lack of appreciation of the fact that the subjective interpretation was adopted (especially by Laplace) *because* it was logically compatible with determinism, and because it was apparently the *only* one that was compatible; Pauli's words even suggest that the subjective interpretation was a kind of anticipation of that allegedly more 'fundamental' way of thinking which we owe to quantum theory and to complementarity. (Indeed it was, but in quite another sense.) In his next sentence, Pauli mentions tendencies to interpret the subjective concept of probability as an irreducible or 'primary' concept of physics; and he ends the paragraph by a reference to what he calls 'the fact that the inductive inferences of the natural sciences are always probability inferences'.

Pauli's subjective interpretation needs, of course, a 'bridge' to allow us to say something objective with its help. His 'bridge' is, it appears, Cantelli's 'strong law of great numbers'.[3] I think it would be fair to translate this law into subjectivist language as follows: 'Assume that *our degree of nescience* about the result of experiment remains constant throughout a long sequence of repetitions of that

[3] *Op. cit.*, pp. 113 f. (Cantelli's law is formulated by Pauli in objectivist language and referred to as 'the theorem of Bernoulli, also called the law of great numbers.')

experiment; and assume that this degree of nescience equals p. (It may be maximal nescience, usually expressed by $p = \frac{1}{2}$.) Under this assumption of *not*-knowing, we can always calculate a large number N with the following property: for any desired degree of certainty short of absolute certainty, we can *know for certain* that the relative frequencies of the results of the experiments will, from the Nth experiment on, deviate as little as we like from our original degree of nescience p.'

This formulation shows, I hope, how absurd it is to believe that we can derive high degrees of certainty from nescience, and that it is strange to suggest that a frequency must equal a degree of nescience. (*Cp. L.Sc.D.*, section 62.) Of course, my formulation differs from the customary one which avoids in cases like this the use of an explicitly subjectivist language; whence originate all the oscillations between the objective and subjective interpretations.

Among Pauli's assertions as quoted here, the one with the greatest interest for us is this: 'If this knowledge of ours is maximal—that is to say, the most precise which is at all compatible with the laws of nature . . . then the entropy is always zero.' This assertion seems to be unchallenged. We find it, for example, in J. von Neumann's famous *Mathematical Foundations of Quantum Mechanics* (1955):[4] 'For a classical observer, who knows all co-ordinates and momenta, the entropy is therefore constant, and is in fact 0, since the Boltzmann "thermodynamical probability" is 1'

I shall not examine here the arguments in support of this thesis because they seem to me unnecessarily complicated; very muddled; highly arbitrary; and somewhat apologetic. Instead I shall first try to contrast it with what I take to be the correct view (I believe it was Boltzmann's), and then to refute it by showing the absurdity of its consequences.

What I take to be the correct view of zero and maximum entropy is this. A system will be in a state of *maximum entropy* if, and only if, the distances of the molecules (and thus the positions) show a chancelike or 'normal' distribution whose mean is equal to a distance in which the molecules are equally spaced over the total available volume, and if, in addition, the velocities (and their

[4]Translated from the German edition by R. T. Beyer; the quotation is from p. 400. Von Neumann refers to L. Szilard's paper in *Zeitschrift für Physik* 53, 1929. See also *L.Sc.D.*, Appendix *iv. [See also *Unended Quest*, 1976, section 36. Ed.]

directions) are distributed in accordance with a chancelike distribution, described by Maxwell's velocity distribution law. In other words, the entropy of a system is maximal if and only if the positions and velocities of the molecules are distributed in a random fashion. If the system consists of very many molecules, this entails that for all its sub-volumes, positions and velocities will also be distributed in a random fashion.

As opposed to this, the system will be in a state of *zero entropy* if and only if it is as far removed from a random state as is at all possible. The kind of state envisaged would be a state in which all or almost all the molecules are packed in a small sub-volume, and in which all or almost all have the same velocity (and direction).

In this way, the two states of extreme entropy can be described objectively.

The description is precisely analogous to the description of a random sequence of zeros (heads) and ones (tails), with equidistribution. Let N be the length of a very long sequence S, and let n be the greatest integer such that 2^{n+1} is smaller than N. We may then call the sequence S *'perfectly random'* if and only if the relative number of the occurrences of any pair, triplet, . . ., n-tuplet, of zeros and ones deviates from that of any other pair, triplet, . . ., n-tuplet by not more than $d = n/N$. (*Cp. L.Sc.D.*, end of section 55, notes *2 and 1.)

Having in this way defined maximum randomness in the case of a sequence—corresponding to its maximum entropy—we can of course easily construct a sequence in which the difference d between the frequencies of certain pairs, triplets, . . . n-tuplets reaches a maximum. This will be the case if, for example, all the *ones* of the equidistributed sequence S are packed into one block—preceded and followed by, say, a block of *zeros*.

The analogy with the state of a gas is obvious. Thus I have shown that we can characterize randomness or entropy in objective terms.

So far I have not mentioned probability. Probability may come in now, however, to tell us that the overwhelming majority of all possible sequences—or of all possible states of gases—will be highly random or entropic. Or in other words, if we arrange an appropriate experiment to produce a sequence of zeros and ones, or a state of a gas, then there will be as a rule a very high propensity to produce random sequences, and gases of approximately maximal entropy.

This tendency or propensity will not, however, become operative if our experimental set-up involves a special method of ordering—for example, a method of producing the result '010101 . . .'.

The overwhelming majority of all other experiments will have a tendency or propensity to produce random states; and accordingly, the overwhelming majority of all states, whether random or not, will likewise have the tendency to be succeeded by random states.

It should be noted that this analysis of random states and random sequences is bound up with the law of large numbers, and therefore only applicable to states of gases which consist of a large number of molecules, or to very long sequences. The question whether a sequence consisting of one or two elements is ordered or disordered is pointless.[5]

On the basis of this analysis it can be shown how the mistaken belief has arisen that *complete knowledge* of the state of a gas would give it the entropy zero. The mistake is connected with the fact that we may start from the assumption that every single possible state of a gas—or every single possible sequence—is equally probable, and that every single state or sequence must therefore have zero (or approximately zero) *probability*. This is correct. It is also correct that high probabilities are in some way closely associated with high entropy, and zero probabilities with zero entropy. From these two correct premises it is then inferred that every single state—if completely given—and every single sequence have zero *entropy*.

But the inference is invalid, because the close association between probability and entropy is not quite as close as all that. It is a fact that every completely given state, or sequence, is equally probable, and hence exceedingly improbable, but this fact simply does not concern us here where we are essentially concerned with a method of *distinguishing* the different states, or sequences, according to their internal structure, rather than with a method which looks upon them as *indistinguishable*—with respect to their probability, that is.

The mistake may also be described thus: we are interested in the *structural properties* of sequences—in the question whether a certain sequence has the *structural property* of disorder (or order). Al-

[5]Von Neumann's problem of applying Boltzmann's ideas to the state of a gas consisting of one molecule seems to be misconceived. His subjectivist solution—that the entropy of a single molecule will depend upon our state of knowledge—stands and falls with his subjective theory. [See *Unended Quest*, 1976, sect. 36. Ed.]

though each sequence itself is equally improbable, it is very probable that it will have the property of disorder rather than of order. There are far more disordered than ordered sequences.

(Or to use an analogous example, we may say that any particular occurrence—for example, that a coin has been tossed here and now—is infinitely improbable. But this does not mean that the property 'heads' of the coin toss is infinitely improbable.)

The belief that any particular given state or sequence has zero entropy is so obviously mistaken that it has hardly ever been held in this blunt fashion. But, as we have seen, it has been used in the subjectivist version. In this form, it was more plausible, because it could be argued that incomplete knowledge may correspond to a whole set of possible states (whose measure might be non-zero) while complete knowledge would correspond to one single possible state whose probability was zero.[6]

Let us now return to our medicine bottle filled with air; and let us assume this time that our knowledge about its microstate is maximal—'that is to say, the most precise knowledge which is at all compatible with the laws of nature'. Then the entropy of the system must be zero, we are told by Pauli and von Neumann.

However, from the objective point of view, the system is (most probably) in a state of disorder or randomness: its entropy will be maximal.

Who is right? From the objectivist point of view, we can easily find out. For having been supplied, according to our assumption, with perfect knowledge, we only need to calculate the distributions of the positions and the velocities and to see whether they are random, or approximately random. We can be pretty certain, *a priori*, that we shall not discover that the system is in a perfectly ordered state. In the highly improbable case that the system is in a perfectly ordered state, the objectivist will admit that he is wrong about this; but he will still insist that this is a highly improbable

[6]There is a further very real difficulty which has contributed to make the subjectivist interpretation of entropy more acceptable. I have in mind the problem of Maxwell's demon: since 1905, the success of the exorcism of the demon has become dubious, and this fact has led to attempts to solve the problem by invoking subjective probabilities, and other measures of subjective knowledge (or 'information'). No doubt some interesting analogies have been unearthed, but the soundness of the edifice, and especially the part played by Maxwell's demon, seems to me highly questionable.

accident which had nothing whatever to do with the fact that we had been invested with perfect knowledge about the system. Moreover, the objectivist will assert that even in the extremely improbable case of our catching the system accidentally in an ordered state we can be pretty certain that it was disordered the moment before and that it will in the next instant return to a state of disorder; and he can test this retrodiction and this prediction by a calculation of one of the next states of the system; a calculation which, having been given perfect knowledge, he will be able to carry out.

All this reasoning is, I believe, consistent with that of Boltzmann, who argued that the system will again and again take up states of order (low entropy), but very rarely, and only for extremely small periods of time which are proportional to the improbability of the state of order, and which decrease very quickly with an increase in the degree of order.

Now let us hear the subjectivist. To him, this reasoning should mean nothing. He has replaced the term 'order' by the term 'knowledge', and the term 'disorder' by 'nescience'. And he may point out that there is a law that partial nescience will quickly increase, and will soon become total nescience, owing to the collisions between the molecules. As Max Born points out, it suffices that we do *not* know the position or momentum of one single molecule: on this assumption our ignorance will soon spread over the whole system, even if we know, to start with, the precise positions and momenta of *all* other molecules. 'Irreversibility', he writes, 'is therefore a consequence of the explicit introduction of ignorance into the fundamental laws.'[7]

This increase of nescience does show a certain similarity to Boltzmann's increase of disorder; but it is sufficiently different to be clearly distinguished.

The fundamental point is that nescience *always* increases, provided we do not start with complete knowledge. But disorder, or entropy, *decreases at times*; according to Boltzmann, it fluctuates. And we have experimental evidence of these 'fluctuations' in the so-called Brownian movement since Einstein's interpretation of 1905. So it does not seem possible to maintain the subjectivist identification of disorder (or entropy) and nescience.

[7] M.Born, *Natural Philosophy of Cause and Chance*, 1949, p. 72. Other passages of an equally subjectivist character may be found on pp. 48 and 59.

Secondly, nescience does *not* increase if we have complete or perfect knowledge to start with (as both von Neumann and Born imply); for perfect knowledge allows us to calculate all future states with perfect precision. Again this is incompatible with Boltzmann's view; for if ever a system should attain perfect order by a highly improbable fluctuation, it would, in all probability, immediately become disordered again, according to Boltzmann.

Thus the often repeated claim that we can, as it were, translate Boltzmann's theory into subjective language must be dismissed.

Having tried to explain what I take to be the correct view, I now turn, as promised, to the refutation of the subjectivist theory of entropy or disorder. I shall attempt very briefly to show that the theory is absurd.

Let us again open our medicine bottle in the vacuum flask. What we wish to explain is why the gas will always escape, and why it escapes irreversibly. Both objectivists and subjectivists explain it as a result of the law that entropy or disorder tends to increase. The objectivist will point out that in the volume now available, the positions of the molecules would be highly ordered if the gas were to remain in the bottle, and that there will be an overwhelming probability, in the sense of an overwhelming propensity, for the gas to assume a state in which the positions of the molecules are distributed in a random fashion over the available space. Thus the objectivist can offer an explanation (never mind whether it is entirely adequate) of why the gas escapes.

The subjectivist, however, cannot offer any such explanation. He cannot even say that the gas has in fact expanded. *All he can say is that his state of ignorance has increased* (provided he did not start with perfect knowledge). This is absurd. For what we wanted to explain was not the state of ignorance, but the state of the gas; not that knowledge has been lost, but that gas has been lost from the medicine bottle.

'The time variations of entropy', von Neumann sums up, 'are then based on the fact that the observer does not know everything.' Thus if the observer knew our bottleful completely, the entropy would not increase. There would be two possibilities: (i) the gas would stay in the bottle, or (ii) the gas would escape. (i) is clearly absurd (it would mean a mystic influence of our knowledge upon

the state of the gas). But if (ii) the gas escapes, its escape cannot be connected with an increase of entropy. Indeed, our perfect knowledge would allow us to predict the escape *without making use* of the law of entropy increase: we could predict the escape by calculating the tracks of all the molecules. But in spite of the fact that we do not need to appeal to the entropy law, it is absurd to deny that the escape proceeds in accordance with it: the gas has escaped, even though, in this particular case, knowledge has not escaped with it. And it was the escape of the gas which was first described as an increase of entropy.

From a determinist point of view, it is indeed difficult to see what statistical mechanics is for: why, if all predictive questions about gases escaping from all single bottles with given initial conditions have been answered, a question should remain; the question, I mean, of the *law-like character*, the *regularity* of the gas escapes, or of the mixing processes, and the question of the *irreversibility* of these processes. These regularities could remain unnoticed, even if we could in every single case predict precisely what will happen.

While I see in statistical mechanics an attempt to explain these regularities, von Neumann describes[8] the statistical treatment as 'a luxury of extra addition' which becomes important only *if our knowledge is deficient*. In this attitude lies the source of the whole trouble: the subjective theory again shows itself as a remnant of a determinist prejudice.

6. *Oscillations Between the Two Interpretations.*

That Pauli wavers between the two interpretations has already been indicated: he reads Cantelli's strong law in an objectivist sense. But von Neumann's and Born's interpretations also oscillate. Thus we find in von Neumann's book[1] a passage in which von Mises's theory of 'collectives' is described as 'in general necessary for establishing probability theory as a theory of frequencies'; and it is implied that the author accepts this theory. But von Mises's views are utterly incompatible with any subjective theory, as von Mises always emphasized; and von Neumann makes no attempt to tell us how the two views can possibly be combined.

[8]Von Neumann, *op. cit.*, p. 298.
[1]Von Neumann, *op. cit.*, p. 206.

Similarly Max Born. I have mentioned his subjectivism; but there are also objectivist passages in Born's work. 'Irreversibility can be understood only by explicitly exempting part of the system from causality' is a passage of this kind.[2] (I feel inclined to agree with the spirit of the passage, even though I do not agree with its wording.) In the continuation of this passage, Born's wavering seems implicit. 'One has to abandon the condition', he writes, '. . . that the positions and velocities of the particles are under control' The term 'under control', which is not explained, seems to me to indicate a wavering between 'are not known'—which would be in keeping with most of his text—and 'cannot be reproduced at will if we repeat the experiment', or in other words 'are not part of' (or perhaps, 'cannot be made part of') 'the objective conditions of the experiment'—which would indicate an inclination towards an objective interpretation.

Such wavering or oscillating is indeed a necessary consequence of any concession to subjectivism. For I do not believe for a minute that the authors cited really *intended* to explain irreversible physical processes, such as mixing, or diffusion, by our ignorance. I suppose that, faced with these consequences, they would withdraw the statements quoted, or at least try to re-interpret them. But I feel certain they took up these positions because they were forced into them by real difficulties inherent in their fundamental approach.

[2]M. Born, *op. cit.*, p. 72.

THE OBJECTIVITY OF QUANTUM THEORY

7. *The Objectivity of Quantum Theory: The Boxes.*

It is often asserted that, in view of the situation in quantum theory, object and subject can no longer be sharply separated.[1] To use Heitler's words, the 'separation of the world into an "objective outside reality", and "us", the self-conscious onlookers, can no longer be maintained. Object and subject become inseparable from each other'.[2] This, according to Bohr, is due to '*the impossibility of any sharp separation between the behaviour of atomic objects and the interaction with the measuring instruments which serve to define the conditions under which the phenomena appear*'.[3] Heitler elaborates the point in some detail. 'One may ask', he writes, 'if it is sufficient to carry out a measurement by a self-registering apparatus or whether the presence of an observer is required.' And he arrives at the conclusion that the self-registering apparatus is insufficient, and that 'the *observer* appears, as a *necessary* part of the whole structure, and in his full capacity as a conscious being'.[4]

We must be grateful to Heitler for giving us what is perhaps the sharpest formulation of this *doctrine of the involvement of the subject with the physical object*; but it also frequently occurs, in one form or other, in Heisenberg's *Physical Principles of the Quantum Theory*, and in many other statements of the orthodox view.

[1] [See the discussion in the Introduction to this volume. Ed.]
[2] W. Heitler, 'The Departure from Classical Thought in Modern Physics', in P. A. Schilpp, ed.: *Albert Einstein: Philosopher-Scientist*, 1949, pp. 194 *f.*
[3] N. Bohr, 'Discussion with Einstein', in *Albert Einstein: Philosopher-Scientist*, *op. cit.*, p. 210. (The italics are Bohr's.)
[4] W. Heitler, *op. cit.* p. 194.

But this doctrine is simply *false*: quantum theory is as objective as any theory can be.

Taking up Heitler's challenge, we first may describe a *classical self-registering experiment* as follows: a box is prepared in which the experimental arrangement is set up, including a self-registering apparatus. Then the box is closed. The reading of the apparatus (within certain limits which may be very narrow) is predicted by the theory. After a time, the box issues a tape or film on which the reading is recorded; and we compare it with the predicted reading.

Now the situation in *quantum theory* is in principle exactly the same, with only one characteristic difference: for every typical quantum theoretical experiment we need, instead of *one* box, *a battery of many boxes*, each fitted with the same experimental set-up—or with set-ups as similar as we can make them; and the theory predicts, in most cases, not the recording of any single tape or film issued by each of the boxes, but the statistical distribution of all these readings produced by all these boxes.

In some special cases, the theory even makes *certain* predictions, i.e., predictions about every single reading issued by every single box. For example, if the experiment is one with a diluted beam of electrons, and the recording apparatus consists of two films, placed in layer fashion on top of each other, which retain 'an image of the point where the electron (or the electrons) passed through',[5] then we shall *always* find that the images on the two films very nearly coincide. We can thus 'predict with certainty the result of this observation'—in direct opposition to the assertion of Heitler[6] whose example is identical to ours, but who arrives at the opposite (and, I am afraid, 'evidently impossible') result; which shows that even a physicist of the first rank can be misled by the quantum orthodoxy.

In Heitler's particular example we do not need more than one box; this is due to the fact that the result in question can be predicted with probability 1 ('with certainty'). But for the more typical quantum experiments, a *battery of boxes* is needed, or a battery of repetitions of the experiment in the one box.

Nowhere in the literature do we find an experiment which cannot

[5]Heitler, *loc. cit.*

[6]*Cf.* Heitler, *loc. cit.*: 'It is then evidently impossible to predict with certainty the result of this observation.'

120

be rearranged, by the method explained here, in 'a self-registering apparatus', or in a battery of apparatuses—in direct contradiction to Heitler's assertion. *Quantum theory is exactly as objective as any other physical theory.*[7]

8. *The Source of the Confusion: the 'Reduction of the Wave Packet'.*

Why then is the theory so often misinterpreted in a subjectivist sense? Why does Heitler believe that the observer plays an essential part in the experiment in which a diluted electron beam impinges upon two films placed in layer fashion on top of one another?

The answer is perfectly clear: the mathematical formalism—that is to say, Schrödinger's wave equation—gives, combined with Born's interpretation, only the probability (density) distribution of the electrons which impinge upon the films. *It does not tell us that the places where the electrons impinge upon the first and the second film will coincide.*

However, this latter result can be obtained if we treat the place of the impact of the electron upon the first film as a new initial condition for a new application of Schrödinger's equation.

The 'observer'—or indeed, the physicist equipped not only with observational results, but with Schrödinger's equation—seems to enter explicitly because the *mathematical formalism* does not yield *in one step* the prediction of the result of the two-film experiment. Two applications of the formalism are needed, and in order to apply the equation for the second time, an *observation* seems to be needed.

This is, in brief, the argument of Heitler (and of the orthodox school). But in so far as it asserts that an observer must intervene in order to make the second step possible—the so-called '*reduction of the wave packet*' (to the point of the first impact)—whenever we apply Schrödinger's equation to *any* co-ordinates of an impact on the first film, it will lead to the prediction that the impact on the second film will have (very nearly) the same co-ordinates. No

[7]This section contains the gist of my opening remarks, as chairman, of the discussion following Professor A. March's lecture on '*Der Gesetzesbegriff in der Physik*', in August 1948, in Alpbach, Austria. (*Cf. Gesetz und Wirklichkeit*, edited by S. Moser, 1949; my remarks were intended to underline and sharpen March's assertions on p. 80.) I did not refer then, of course, to the passages from Heitler and Bohr, which were not yet published.

observation is needed for this result—neither an 'actual' nor a 'potential' one.

This is why the argument fails, and why our self-registering boxes work, in the sense that we can make successful predictions, not only for the statistics of those results which exhibit scattering, but also for those results—like the *coincidence* on our two superimposed films—which do not show any further scatter.

The orthodox argument does establish, however, though unintentionally, the incompleteness of the mathematical formalism.

9. *The Incompleteness of the Mathematical Formalism of Quantum Theory.*

There are many senses in which one can speak of the completeness or incompleteness of a theory, or of a mathematical formalism. The idea of the completeness of a mathematical formalism which I shall use here is intended to cover a certain range of physical phenomena in the following sense: the formalism is *complete* if and only if it covers 'automatically' (i.e., without the need to make a non-formal 'second step', in the sense of the preceding section) any prediction of experimental results which we can make in the range of phenomena in question.

The proposal to use the term 'complete' in this sense comes from Niels Bohr. In his reply to a paper by Einstein, Podolsky, and Rosen entitled 'Can Quantum-Mechanical Description of Physical Reality be Considered Complete?'[1], Bohr wrote: 'Such an argumentation, however, would hardly seem suited to affect the soundness of quantum mechanical description, which is based on a coherent mathematical formalism *covering automatically any procedure of measurement* like that indicated.'[2] Since the 'soundness' in question was the problem of completeness, and since by a 'procedure of measurement', covered by the formalism, Bohr clearly means experimental results predictable by the formalism, one might say that the definition of 'complete' here suggested is, as near as it can be, Bohr's.

But according to this definition, the formalism is *incomplete*—in

[1] *Physical Review* **47**, 1935, pp. 777–80.
[2] Niels Bohr, *Physical Review* **48**, 1935, p. 696. (The italics are mine.)

direct opposition to Bohr's contention. In fact the orthodox school tries to have it both ways.

Bohr says the formalism covers all cases automatically. But Heitler says that the observer must intervene; or as Heisenberg put the same point: 'The discontinuous "reduction of the wave packets" *which cannot be derived from Schrödinger's equation*, is . . . a consequence of the transition from the possible to the actual.'[3] But it is certainly not possible to insist on the one hand that the formalism is complete and to insist on the other hand that its application to 'the actual' actually demands a step *which cannot be derived from it.*

In order not to be misunderstood, I wish to stress that I do not think that the incompleteness mentioned here is a serious one. I do not even suggest that it is the one which Einstein had in mind. I shall show that, in spite of this incompleteness in the formalism, there is *no incompleteness in the interpreted theory*; indeed, this particular incompleteness in the formalism is one which is characteristic of the probabilistic treatment of a wide range of problems which have nothing to do with quantum physics or interference by observers or anything of that kind. For this very reason, there is no need to remedy this particular incompleteness.[4] All I wanted to show was that the formalism of the theory is incomplete, precisely in Bohr's sense, as against Bohr's assertion; and that it is just this typical and harmless incompleteness which has led to the myth of the observer's entry into quantum theory. [See the discussion of completeness in the 'Preface 1982'.]

10. *A Random Walk and the 'Transition from the Possible to the Actual'.*

A soldier in the desert is given a pocket roulette-wheel and instructed to set it up, letting the pointer spin and come to rest, and then to march for one minute in the direction indicated by the pointer; and then to repeat the performance, letting the pointer spin again, and again marching for one minute in the direction of the pointer. It is at once intuitively clear that, on the basis of this

[3] W. Heisenberg, in *Niels Bohr and the Development of Atomic Physics*, 1955, p. 27. (The italics are mine.)

[4] As has been attempted, for example, by L. Jánossy, *Annalen der Physik* (sixth series) 11, 1952, pp. 324 *ff.*

instruction, we obtain a probability distribution for the position of the soldier which spreads with the soldier's marching speed from the starting point in all directions—a kind of cloud which is dense in the centre and thins out towards its periphery. (There is a kind of Huygens principle valid for this cloud: every potential end of one minute's walk becomes the centre of a new cloud which is superimposed upon the others.)

Now let us *observe* the soldier, after one hour, when he is just consulting his pocket roulette-wheel. Then, we may say, the old cloud disappears, and a new cloud is started at the point at which we have observed the soldier.

This is precisely the same thing as the 'reduction of the wave packet', and it occurs, precisely in this way, in every so-called Markov-chain problem of probability theory, and, slightly more generally, in every case in which *relative* probabilities play an essential part. It even occurs, though in a slightly more trivial way, in any such game as tossing pennies: relative to the information that we have tossed heads, the (relative) probability that we have tossed heads changes from ½ to 1. That this is fundamentally the same problem as that which has disturbed the quantum theorists (that it is therefore the same as Heitler's problem, and the same as that which makes Heisenberg talk about the 'transition from the possible [or potential] to the actual') I asserted many years ago, especially in *L.Sc.D.*, section 76.

The harmlessness of this famous 'reduction of the wave packet' is, of course, why I consider the incompleteness of the formalism to be harmless: once we can interpret it properly, the *interpreted theory*—rather than the mathematical formalism as such—is in this particular point as complete as we can wish it to be.

In all these cases the 'reduction of the wave packet' has of course nothing whatever to do with an alleged interference of the observer, or of the observing instruments, with the objects observed: soldiers, and even pennies, are too heavy to be upset by the few light quanta needed for observing them. This is finally admitted, implicitly, by Heisenberg, in the paper quoted, for he now finds (pp. 25 and 27) the 'transition from the possible to the actual' also in classical thermodynamics. But this destroys the orthodox (or 'Copenhagen') interpretation which he defends in that paper. For the whole issue is whether quantum theory is so very different from other probabilis-

tic theories as the orthodox school always asserted; whether in quantum theory, as opposed to 'classical' or 'macroscopic' theories, we are faced, owing to the peculiarity of the object dealt with and owing to Planck's constant h, with a new and unique situation.

11. Particles, Waves, and the Propensity Interpretation.

The problem of particles vs. waves, and of the relations between particles and waves, has for a considerable time been discussed by some of the founders of quantum theory (especially by Bohr, Born, Schrödinger, Heisenberg, Dirac, and later by Landé). In these discussions, a certain element of irritation or exasperation is sometimes discernible, directed especially against Schrödinger who, it is indicated, has 'a personal attachment to the waves', and 'is in love with this idea'.[1] The irritation is perhaps even more marked among physicists of the younger generation who feel inclined to look upon the problem as a philosophical pseudo-problem rather than a physical one. For it is a problem of neither (i) the *mathematical formalism* nor (ii) its *application*. On both of these points, the contesting parties agree.[2] Thus there does not seem to be a problem here that need be taken seriously by the tougher physicists.

Those tough, no-nonsense instrumentalists who think like this should pause to consider that it was Schrödinger who made the discovery of the mathematical equivalence of the two theories; who first developed the 'transformation theory' (of vectors in Hilbert

[1]Cf. Max Born, 'The Interpretation of Quantum Mechanics', *British Journal for the Philosophy of Science* **4**, 1953, p. 98.

[2]This is also stressed in Born, *op. cit.*, and in Heisenberg's article, quoted in note 3 to section 9 above. Both criticize Schrödinger: Born (on p. 98, footnote), for relying on the intuitively understandable 3-dimensional waves which result from the 'method of second quantization' (due to Jordan and Klein, *Zeitschrift für Physik* **45**, 1927, p. 751, and Jordan and Wigner, *ibid.* **47**, 1928, p. 631; see Schrödinger's contribution to *Louis de Broglie, Physicien et Penseur*, 1952), and in which, Born says, 'the statistical character of the ψ-function is introduced on an even deeper and more abstract level'. In the second article, Schrödinger is criticized by Heisenberg (p. 24) for his 'misunderstanding of the usual' (orthodox) 'interpretation': a misunderstanding which consists in 'overlooking the fact' that in the orthodox interpretation, the 3-dimensional waves of Klein, Jordan, and Wigner 'have no direct connection with probability waves', i.e., are not of a statistical character, but 'have a continuous density of energy and momentum, like a Maxwell field'. Thus the two defenders of the orthodox view are here seen to agree only in one point: that Schrödinger must be wrong. But their arguments contradict each other. (See also note 1 to section 28 below.)

space[3]), and who wrote as long ago as 1926 about his equivalence proof: 'On this view' (that of Kirchhoff and Mach) 'mathematical equivalence has almost the same meaning as physical equivalence.' Thus if Schrödinger believes that the two theories, that of waves and that of particles, are not equivalent physically, the reason is not a failure to see a point which, in fact, nobody saw as early or, I believe, as clearly as he did.

In fact, Schrödinger was too generous when he conceded as much as 'almost' physical equivalence: his wave theory was a theory of both free and bound electrons, which the particle theory was not, at least at the time.[4]

If Schrödinger's attempt to criticize the orthodox view led to such replies, my views, which are merely those of a philosopher, have little hope of being taken seriously by physicists; the hope is even lessened by the fact that, although I am deeply in love with the waves (much more than with the particles[5]) I nevertheless believe that the quantum theory is in a very definite sense a particle theory (here I disagree with Schrödinger) and in a sense which excludes a duality, or analogy, or complementarity, between particles and waves. To be more explicit, I believe that the waves (even those of the second quantization) are mathematical representations of *propensities*, or of dispositional properties, of the physical situation (such as the experimental set-up), interpretable as propensities of the *particles* to take up certain states. In so far as quantum theory is a probabilistic theory, it is a theory of waves, or of matrices: this is why the wave theory and the matrix theory are equivalent. But the probabilities which the theory determines are always the propensities of *particles* to take up a certain state under certain conditions.

[3]*Cf.* Schrödinger's *Collected Papers on Wave Mechanics*, 1928, pp. 45–61, esp. footnote on p. 52.

[4]The papers by M. Born and N. Wiener, *Journ. Math. and Phys.* 5, 1926, p. 84, and *Zeitschr. f. Phys.* 36, 1926, p. 174, operated with 'continuous matrices', no doubt an *ad hoc* adaptation of the matrix method to free particles. Schrödinger's theory, on the other hand, developed de Broglie's idea of *explaining* (not *ad hoc*) the Bohr stationary states by wave-properties also possessed by free particles, and independently testable in their case.

[5]The reason is that the wave (or field) theory offers a possibility of *explaining* matter by something that is not matter (and more general than matter); for further reasons, see the preceding footnote. [See Popper's discussion of materialism and its history in *The Self and Its Brain*, 1977, especially sections 3–5, and Chapters P3 and P5. See also his 'Philosophy and Physics', *op. cit.*, and the Metaphysical Epilogue below. Ed.]

There is thus no symmetry or duality between particles and waves: the waves describe dispositional *properties* of the particles. As Alfred Landé puts it: 'If this . . . is the famous *duality*, then swans must be . . . dualistic birds, since on the one hand they *are* individuals or particles; on the other hand, they *have* the quality of wavy necks; so that "one will for ever remain uncertain whether swans are really particles or waves". This duality is a contrast of incontrastibles.'[6] I fully agree, except that I would have preferred a more obviously dispositional property than wavy necks;[7] for example, the swans' disposition to make, under certain circumstances, wavy movements with their necks.

According to this view, the particles do not appear, explicitly, in the formalism—neither in the wave formalism, nor in the matrix formalism (which is often said to be a particle formalism), nor yet in the 'neutral' transformation theory. All these theories describe the propensities of certain variables to take up certain values; variables which may be *interpreted* as variables of the state of certain particles.

One of my worries is that this view would not be acceptable to Schrödinger. Yet it is really not far removed from his own view. One must remember that, according to the interpretation proposed here, propensities, and fields of propensities, are just as real as forces, and fields of forces. They are, like forces, dispositional properties; and like forces, or fields of forces, they are properties not so much of the particles as of the total physical situation; they are, like forces, *relational* properties.

It may be worth while to explain this point a little more fully. One reason why Newton found it difficult to accept his own theory as an *explanation* of gravity was that he found it difficult to see in his attracting forces—acting at a distance—an *inherent* or an *essential* property of matter. Extension could be an essential property of a piece of matter, because it did not depend upon anything besides this piece of matter. But attraction was a disposition to attract *other pieces of matter*: it was a relational property, a mutual property; and

[6]A. Landé, 'The Logic of Quanta', *British Journal for the Philosophy of Science* 6, 1956, p. 316.
[7]In fact '*all* universals are dispositional', as I have stressed on several occasions. (See for example my paper, 'Three Views Concerning Human Knowledge', *Contemporary British Philosophy* 3, 1956, section vi, near the end, reprinted in *Conjectures and Refutations*, 1963; see also *L.Sc.D.*, end of section 25; and *Realism and the Aim of Science*, Vol. I of the *Postscript*, Part 1, section 11.) Yet some are more dispositional than others.

it was therefore difficult to conceive it as *inherent* in the piece of matter—as difficult as it is to regard as inherent the property in a piece of matter of being *more extended (bigger) than another piece of matter.* [8] It is our Aristotelian and essentialist habit of expressing ourselves (connected with a habitual preference for using 'categorical', i.e., subject-predicate statements) which suggests that relational properties are not 'real' but 'ideal'; that is to say, that they are founded only in our own thinking and ordering activities. It is this attitude, in the last analysis, which leads to the subjective interpretation of probabilities, and which makes it difficult to accept an objective propensity interpretation.

Once we allow that measures of possibilities may be interpreted as propensities to react in certain ways under certain given circumstances (initial and boundary conditions), our interpretation in terms of possibilities is no longer opposed to the interpretation which Schrödinger proposed in 1926, in his fourth paper on 'Quantization as a Problem of Proper Values'. [9] The difference between them, if any, is connected with the *fact* that no propensity theory of probability existed at the time, and with the *view* that a singular probability theory cannot be objective, and that an objective theory of probability must be a *purely* statistical or frequency theory. Instead of speaking of 'measures of possibilities' and of a 'measure-function' (which determines the density distribution of the measure of probability), Schrödinger speaks in this paper of a 'weight' of the possibilities—i.e., of all 'point-mechanical configurations [which are] kinematically possible'—and of a 'weight function'. He writes: 'we may say that the system exists, as it were, simultaneously in all the positions (which are) kinematically imaginable, but not "equally strongly" in all.' How strongly it exists in each of these positions is determined by a weight function, i.e., by the square of the wave amplitude.

Schrödinger comments: 'This new interpretation may shock us at first glance, since previously we have often spoken of "ψ-vibrations" in such an intuitively concrete way as though of something quite real. But there is something tangibly real behind the

[8] See also my paper 'Three Views Concerning Human Knowledge', *op. cit.*, especially the paragraph to which the seventh footnote of section iii is appended.
[9] *Annalen der Physik* (fourth series) **81**, 1926, p. 135; *Collected Papers on Wave Mechanics*, pp. 120 f.

present conception also'[10] No doubt this view does have much in common with the propensity interpretation. Schrödinger emphasizes the objectivity or 'reality' of the waves—the fact that waves can interact with each other as well as with other physical magnitudes; and at the same time, he emphasizes the fact that they are somewhat less real and less concrete and intuitive[11] than ordinary waves or vibrations.

So far there is agreement with the propensity interpretation. The difference is this: Schrödinger does not interpret his weight as a tendency (or disposition or propensity) to realize itself which is attached to a possible configuration, or to a possible kinematic state; a tendency whose strength or weight expresses itself by the statistical frequency with which the state in question occurs in a sequence of repetitions of the experiment in question. In other words, Schrödinger does not interpret his weight as a *probability*, although he realizes that it *can* be so interpreted. In a paper[12] written one year after the one just quoted, he writes: 'If the reader likes, however, he can understand everything that has been said . . . in accordance with Born's theory,[13] in which . . . the squares of the amplitudes are not interpreted as simultaneous excitation strengths in the single system, but merely as probabilities (relative frequencies of occurrence) of the discrete quantum states in a virtual aggregate.' The 'simultaneous excitation strengths in the single system' are the same, of course, as the 'weights' of the older paper; and one sees clearly that Schrödinger wishes to oppose here an objective property of the *single* system—an intensity distribution—to a relative frequency of occurrence in an aggregate of systems.

The propensity interpretation bridges this gap. It allows the *single system* to have objective tendencies or propensities of varying strengths or weights or intensities, and shows that, if these are interpreted as tendencies to realize themselves, they lead to corre-

[10]*Loc. cit.* (I have transposed a few words of the translation.)

[11]Schrödinger stresses, *loc. cit.*, 'the fact that the ψ-function cannot and may not be interpreted directly in terms of three-dimensional space . . . because it is in general a function in configuration space, not real space'; yet he is blamed 26 years later by Born, *op. cit.*, for failing to realize this fact.

[12]'The Exchange of Energy according to Wave Mechanics', 1927; see *Collected Papers*, p. 146.

[13]M. Born, *Zeitschr. f. Phys.* 37, p. 863; 38, p. 803; 40, p. 167, 1926. (This footnote is Schrödinger's.)

sponding statistical frequencies in a virtual (or real) aggregate or collective.

Schrödinger remarked that 'if the reader likes he can understand everything' in the statistical sense. This indicates, I believe, that he saw, and admitted, a connection with statistics, but wished to stress that he was not satisfied by a *purely statistical* (as opposed to an objective singular) interpretation. We may thus claim, perhaps, that what we call the propensity interpretation fulfils all the intentions of Schrödinger—and perhaps even those of Born, in spite of their apparent conflict.

12. *Partial Anticipations of the Propensity Interpretation.*

Schrödinger had then stressed very early one most important aspect of the propensity interpretation—the objectivity and reality of the waves (in configuration space), and the fact that the wave amplitudes provide a weight, or strength, or measure, of the possibilities.

Born, on the other hand, stressed the opposite, the statistical aspect of our interpretation. But he too came close to the complete propensity interpretation in a remark which may be compared with Schrödinger's comment as quoted in the foregoing section (that 'if the reader likes . . . he can understand everything . . . in accordance with Born's theory'). For Born writes: 'The question . . . whether the waves are something "real" or a fiction to describe and predict phenomena in a convenient way is a matter of taste.'[1]

Born continues: 'I personally like to regard a probability wave, even in 3N-dimensional space, as *a real thing, certainly . . . more than a tool for mathematical calculations.* For it has the character of an invariant of observation; that means, it predicts the results of counting experiments, and we expect to find the same average numbers, the same mean deviations, etc. if we actually perform the experiment many times under the same experimental condition.'[2] The whole passage reads like a manifesto of the propensity in-

[1]See *Natural Philosophy of Cause and Chance*, p. 105. This remark, though not an outright declaration of instrumentalism, contains a big concession to it, to say the least.

[2]The italicized phrase (the italics are mine) is clearly anti-instrumentalist in its tendency, as opposed to the commencing remark. The whole passage is from *op. cit.*, pp. 105 *f.*

terpretation—in strangest contrast to the continuous wavering, or oscillating, between a subjective and a purely statistical interpretation of probability, to be found in the same book. (See above, section 6.) The passage just quoted contrasts so strangely with the extensive use made of the *subjective* interpretation for 'solving' the conceptual difficulties of classical statistical mechanics that one feels inclined to believe, at this stage, that Born is at heart defending an objective propensity interpretation in quantum theory only, but not in classical statistics. But this is not so; for in the immediate continuation of the passage, Born surprises us again by arguing in favour of a quite general propensity interpretation of probability: 'Quite generally', he writes, 'how could we rely on probability predictions if by this notion we do not refer to something real and objective? This consideration applies just as much to the classical distribution function . . . as to the quantum-mechanical density matrix'

This looks like a perfectly clear statement of the propensity interpretation, especially when one finds that, in the sequel, Born discusses Bohr's two-slit experiment (which, I believe, is indeed one of the strongest arguments for the propensity interpretation) in an exemplary fashion to illustrate and to support his theory.

And yet, one can hardly say that Born has reached anything like the position which I have labelled the 'propensity interpretation' either of probability theory or of quantum theory. Admittedly, he reached it in this one passage, but he never realized that this interpretation should be given a good trial; that it is worth an attempt to carry through *consistently*; rather, he takes it as one of the many 'complementary' ways in which the situation may be looked at, for some purpose or other. This explains the concession made to instrumentalism in Born's opening remark, quoted above. It explains the otherwise incomprehensible fact that in the same book in which this eloquent plea is made in favour of a completely general objective interpretation of probability, statistical mechanics is developed and elaborately defended as a purely subjective theory. And it explains how, three years after the publication of this book, Born could write about 'the statistical interpretation of quantum mechanics' as follows: 'This is not a simple matter, and demands not only knowledge of a complicated mathematical formalism, but a certain philosophical attitude: the willingness to sacrifice traditional

concepts and to accept new ones, like Bohr's principle of complementarity.' Now as to 'the knowledge of a complicated mathematical formalism', it must be said that if Born's eloquent plea for a completely general objective interpretation of probability is taken seriously, this cannot be upheld; for in this case, the problem of interpreting probability in such games as tossing pennies would be fundamentally the same as that of interpreting probability in quantum theory. Nor would there be any need to appeal to Bohr's principle of complementarity.

Prominent among others who partly anticipated the propensity interpretation is Dirac. Jeans should also be mentioned.[3]

A clear formulation of the propensity interpretation is also found in a paper of Heisenberg's; and surprisingly, Heisenberg claims that it is the orthodox view. But it soon turns out that Heisenberg does not use the interpretation which he so clearly describes—any more than Max Born does. He really thinks that it is the same as the old orthodox complementarity view, and thus falls back upon his old doctrine of the perfect duality of particles and waves, of quantum jumps (for which the reduction of wave packets is taken as an example), the role of the observer, and all those other orthodox ghosts whose exorcism is one of the minor results of the propensity interpretation, if fully understood.

In this paper, Heisenberg attributes the introduction of what I call the propensity interpretation to the famous paper of Bohr, Kramers, and Slater (1924) which was a precursor of quantum mechanics: ' . . . the . . . interpretation . . . by Bohr, Kramers, and Slater', Heisenberg writes, ' . . . contained some very important features of the later, correct, interpretation. The most important of these was the introduction of the probability as a new kind of "objective" physical reality. This probability concept is closely related to the concept of the natural philosophy of the ancients such as Aristotle; it

[3]Sir James Jeans, *The New Background of Science*, 2nd edition, 1934. On pp. 229 to 233 there is the beginning, as it were, of a most promising discussion of subjective *vs.* objective probability; but although highly instructive, the discussion does not lead to any positive proposal. This is recognized on p. 241 where the discussion is summed up by saying: 'We there described the waves of the wave picture as "waves of probability", but were unable to assign any precise meaning to the term.' In the sequel, Jeans adopts the subjective interpretation ('our lack of precise knowledge', p. 242).

132

is, to a certain extent, a transformation of the old *"potentia"* concept from a qualitative to a quantitative idea.'[4]

The ascription of a new idea of objective probability of this kind—probability as a measure of a potentiality or disposition or propensity—to the paper by Bohr, Kramers and Slater is surprising, but perhaps defensible, even though the idea is at best only implicit in their approach. But I found it even more surprising to see this idea described as one of the 'very important features of the later, correct, interpretation', which Heisenberg identifies with the 'Copenhagen Interpretation', worked out in 1927 by himself and Bohr (*op. cit.*, p. 15); at least I found no trace of this idea in any of Bohr's or Heisenberg's early papers; and this is no accident. For in the orthodox Copenhagen interpretation, ideas such as that of the part played by observation and measurement are of decisive importance, and the indeterminacy relations, which forbid precise measurements, have the main function of introducing into the theory an element of uncertainty, and with it, probability considerations. In other words, probability theory enters because of certain limitations which are imposed by the theory upon our measurements, and therefore upon the precision of *our knowledge*—which gives the theory its *characteristic mixture of objective and subjective elements.* All this appeared suddenly to have been thrown out, and a purely objective potentiality or propensity interpretation appeared to have been put in its place.

Yet there was no cause for alarm. In spite of the passage quoted, there was really no change in the orthodox theory. Exactly as in the case of Born, the real potentialities of the Aristotelian *potentia*, i.e., of our propensity, are not realized in Heisenberg's paper. Heisenberg still operates with the quantum mysteries, the 'three articles of faith' as Landé calls them[5]—duality, uncertainty, and complementarity; and, in addition, with all those other gratuitous quantum mysteries and horrors such as subject-object confusion, the 'transitions from the possible to the actual' as a consequence of our measurements, and quantum jumps.

Among those who anticipated the propensity interpretation, only

[4]See W. Heisenberg, 'The Development of the Interpretation of the Quantum Theory', in *Niels Bohr and the Development of Physics*, 1955, pp. 12 f.

[5]'The Logic of Quanta', *British Journal for the Philosophy of Science* 6, 1956, p. 301.

Landé tried to apply it systematically and consistently in order to dispel most of the famous but gratuitous quantum mysteries with its help. I think that the priority for this interpretation belongs largely to him.[6] He first establishes the need for an objective and singular probability (see *The Open Universe: An Argument for Indeterminism*, Vol. II of this *Postscript*); recognizes that this step alone implies the rejection of determinism; bases the theory upon relative or transition probabilities;[7] and, most important, tries to *explain*, on this basis, the quantum rules which, as he points out, have become articles of faith.

There is only one point of importance where I feel that he does not fully exploit the power of the propensity interpretation. It is his attempt to explain 'the symmetric . . . and the antisymmetric . . . modes of interaction' by the '*indiscernibility of identical particles* in every observational respect'.[8] This explanation is not satisfactory; nor is there any need for it. The appeal to Leibniz's principle is hardly justified since Leibniz does not refer to *observational* indiscernibles; on the other hand, the positivistic principle of the identity of *observational* indiscernibles seems to me untenable, not to say repugnant.[9]

But the main point is that a resolute application of the propensity interpretation makes the appeal to a separate principle unnecessary. For Pauli's exclusion principle may be incorporated as part of the wave theory of the electron in the form of the postulate that of the three possible symmetry classes, only the antisymmetrical class

[6][For Landé's work, see his: *Foundations of Quantum Theory*, 1955; *From Dualism to Unity in Quantum Physics*, 1960; and *New Foundations of Quantum Mechanics*, 1965. Ed.]

[7]These are the Markov chains upon which Fényes bases his most elegant development; *cf.* I. Fényes, *Zeitschrift für Physik* 132, 1952, pp. 81–106; see esp. pp. 103–5. There it is shown that the quantum theory, i.e., the Schrödinger equation, is deducible from statistical diffusion theory *in conjunction with* a classical variation principle—the Euler-Lagrangean differential equation. (The physical meaning of this conjunction is still in need of careful discussion.)

[8]Landé, 'The Logic of Quanta', *op. cit.*, pp. 319 (3), 313; see also his *Foundations of Quantum Theory*, p. 65.

[9]See my paper 'Language and the Body-Mind Problem', *Proceedings of the XIth International Congress of Philosophy* 7, 1953, pp. 101 *ff.* [Reprinted in *Conjectures and Refutations*, 1963. Ed.] As indicated there, if two banknotes are observationally indistinguishable (even in their numbering) one of them may be genuine but we have every reason to believe that one of them must be a fake; and a fake does not become genuine because it is a perfect fake.

occurs.[10] In this form, it does not refer to indistinguishable particles. And from the point of view of the propensity interpretation, it assures that the probability is zero of finding a system that violates the various corollaries (such as the one establishing the Fermi-Dirac statistics) of the exclusion principle.

13. *Are there Quantum Jumps?*

The title of this section is taken from a paper by Schrödinger[1] in which, I believe, he shows that quantum theory does not warrant an interpretation according to which there are quantum jumps.

Quantum jumps were first introduced by Bohr with his atom model of 1913. They were sudden transitions of electrons from one stationary state to another, transitions which were then said not to take up any time. However, the emission of a light quantum (which the theory connects with the transition) does take up time, according both to the new quantum theory and to experiments. From the point of view of the theory, the fact that the emission takes time can be seen, for example, from the indeterminacy relations, since otherwise the frequency (energy) of the quantum would have to be indeterminate. From the point of view of experiment, the average time can be calculated, for example, from the average length of a coherent wave train (3 to 4 feet in length)[2].

Quantum jumps are an old aversion of Schrödinger's. Personally I do not feel any distaste for them—I should be perfectly happy to accept them if physicists could show good reasons for doing so. But I no longer believe that there are such reasons; at least none are given in the two (somewhat contradictory) replies which Schrödinger's

[10]See for example W. Pauli, 'Exclusion Principle and Quantum Mechanics', *Les Prix Nobel en 1946*, esp. pp. 135 f. Pauli justly expresses there on p. 136 his disappointment at the fact that the new quantum theory did not succeed in explaining (or deducing) the exclusion principle. But the form in which it has to be assumed seems to me less *ad hoc* by far than Landé's application of what he calls 'Leibniz's principle'.

[1]E. Schrödinger, 'Are There Quantum Jumps?', *British Journal for the Philosophy of Science* 3, 1952, pp. 109–23 and 233–42.

[2]Schrödinger, *op. cit.*, p. 113. Implicitly, all this was already indicated by Planck in 1923 (Preface of the 5th edition of his *Theorie der Wärmestrahlung*), and admitted by Bohr in 1927; *cp. Atomic Theory and the Description of Nature*, 1934, pp. 81, 83 f., 86 f. See also Introduction to the present volume, note 38a.

paper has received, one from Max Born,[3] and a later one from Heisenberg.[4] Born, so far as I can see, does not answer Schrödinger's question at all. Instead, he discusses a completely different question which he introduces in the title of his section 2: *'Are there Atoms?'*. He claims that Schrödinger's real intention is to deny that atoms exist. This seems to be an inexplicable misinterpretation of Schrödinger's views, and more especially of what he says in the place criticized by Born. Schrödinger's theory may explain atoms (not as waves, but as wave *structures*); but it does not explain them away.

While no answer to Schrödinger's question is given in Born's paper,[5] Heisenberg does give an answer. At first, he does not seem to say more than Born, for he appeals (instead of to the existence of atoms) to the existence of an 'element of discontinuity . . . which is found everywhere in atomic physics (very obviously, for instance on the scintillation screen)'.[6] This clearly does not carry us one step beyond Born's position. That scintillations are news to Schrödinger is news to me; but it really seems that it is no longer generally known that Schrödinger never denied these discontinuities, but that he tried to *explain* them—and not entirely without success. But after this, Heisenberg makes a definite suggestion concerning our question; he says: 'In the usual interpretation of quantum theory it' ('it' is the discontinuity, and presumably the quantum jump) 'is contained in the transition from the possible to the actual.' Having thus made his own counter proposal to Schrödinger's proposal to abandon the myth of quantum jumps, Heisenberg concludes his criticism of Schrödinger—of the man who explained stationary states and quantum numbers—by saying: 'Schrödinger himself makes no counter proposal as to how he intends to introduce the element of discontinuity, everywhere observable, in a different manner from the usual interpretation.' (Were eigenvalues no proposal?)

Thus Schrödinger receives two answers: the first alleges that his denial of quantum jumps is equivalent to a denial of atoms or other

[3]M. Born, 'The Interpretation of Quantum Mechanics', *Brit. Journ. Phil. Sc.* **4**, pp. 95–106 (quoted above).

[4]Heisenberg, *op. cit.*

[5]I neglect Born's discussion of *collisions, op. cit.*, p. 104; although it appears (as Dr J. Agassi pointed out to me) that in confining his discussion to the stationary theory of collisions, he misses Schrödinger's point.

[6]Heisenberg, *op. cit.*, p. 24. *Cp.* also pp. 26 *f.* for the 'reductions of the wave packet'.

'discontinuities'; and the second suggests that the quantum jumps are identical with the famous, or rather notorious, 'transition from the possible to the actual'—that is to say, with the 'reduction of the wave packet'.

I discussed this reduction long ago, in *L.Sc.D.*, section 75, and have now done so again, in section 8 of this volume. No doubt, the reduction of a wave packet can happen very suddenly; even with super-luminal velocity, as I explained in section 75 of *L.Sc.D.*; for it simply is not a physical event—it is the result of the free choice of new initial conditions (or of using a new b in '$p(a,b)$').

What is most remarkable is that Heisenberg himself, in the same paper, admits that this 'reduction' is not a quantum effect; for he says (p. 27) that it is 'exactly as in Gibbs' thermodynamics, a consequence of the transition from the potential to the actual'. Thus his answer to Schrödinger amounts to the assertion that there *are* quantum jumps—exactly as there are quantum jumps in the classical theory.

These answers strongly suggest that there really appears to be no evidence for quantum jumps. But if this is so, why not say so? And why attack Schrödinger when he says so?

If we now look at the problem from the point of view of the propensity interpretation, it becomes, perhaps, a little clearer at one point. First, it is clear that there cannot be quantum jumps in the old sense, because of the indeterminacy principle (with respect to time and energy). Secondly, the 'reduction of the wave packet' is certainly no quantum jump, whatever Heisenberg may say; for otherwise soldiers and pennies would make quantum jumps. Thirdly—and this is the result of the propensity interpretation— quantum theory is not a theory which describes dynamical processes in time, but it is a probabilistic propensity theory which gives weight to various possibilities. Thus even where it does not imply that a process takes time, the process *may* nevertheless take time. I have in mind the case of interaction between a particle and a polarizer; it appears that the particle 'jumps', in consequence of this interaction, into one of two states, becoming polarized either perpendicular or parallel to the optical axis of the polarizer. This is suggested, not only by Dirac,[7] for example, but also by Landé.[8]

Now this may be true, for all I know (although I do not believe

[7]Dirac, *The Principles of Quantum Theory*, 3rd ed., 1947, p. 7.
[8]Landé, *Foundations of Quantum Theory*, p. 18.

that it is). But it seems to me fairly clear that nobody knows it to be true, or even to be a consequence of the theory. All the theory tells us—in its propensity interpretation—is that there are certain propensities for the particles to take up one or the other of two states upon interaction. It even tells us that comparable processes do take time (such as interaction with a light-absorbing filter, for reasons analogous to those connected with the emission of a light quantum).

Thus it seems to me clear that in this case too, *the theory does not imply that there are quantum jumps*; and I think this is what Schrödinger said; though he not only said it very much better, but made many other most interesting suggestions for the further development of the theory.

14. *Are there Particles?*

We have seen that Born and Heisenberg counter Schrödinger's pertinent question 'Are there quantum jumps?' by suggesting that the question which Schrödinger really wanted to ask was, 'Are there atoms?' or 'Are there discontinuities?'. I have tried to show that these suggestions are somewhat evasive, and hardly fair to Schrödinger.

But there is no reason why the suggested question should not be taken up. Are there particles? Are there atoms, electrons, photons, nucleons, and the host of other particles which have been discovered during the last twenty-five years? The answer has to be a *qualified* 'yes'.

To make clear what is meant by a *qualified* 'yes', we have first to distinguish two different questions. The first is the question whether one scientist or another believes, or conjectures, that there are particles. The second is whether the existence of particles is implied in the *interpretation* of the present quantum theoretical formalism.

To the first question the answer is that Schrödinger no doubt believes in particles, but that he was the first among the quantum theorists to see both the possibility of explaining particles by something else, and the need for such an explanation. He did so at a time when the 'elementary particles', as the electrons and protons were usually called, were still generally taken to be indestructible bricks. Today, we no longer use the word 'particle' with the

implication that these things are indestructible. *But it should not be forgotten that the first suggestion that particles may have a limited life came from Schrödinger's theory.*

Nobody, least of all Schrödinger, would suggest that his theory was successful in every respect. But his fundamental position—that particles are not ultimate but are *in need of an explanation,* in terms of something other than particles—will hardly be questioned.

Schrödinger's theory was the first and, fundamentally, the only attempt in this direction, and one might say that the new meaning of the term 'particle' owes very much, if not everything, to the fact that we have learned from his theory how to think in terms of particles which are not ultimate or 'elementary'. In other words, although it does not seem that his theory provides a satisfactory explanation of particles and their transitions, it has shown that, and how, an explanation might be given. Indeed, this is all we possess by way of an approach to theory in this field.

All this should be remembered if we wish to assess Heisenberg's remark that 'Schrödinger himself makes no counter proposal as to how he intends to introduce the element of discontinuity'—for example the discontinuity due to the existence of particles. It should also be remembered if we wish to assess Pauli's comments on the 'opposition' against the orthodox school. 'There is no lack of an opposition', Pauli writes,'. . . but it has remained barren and got stuck in a state of regressive [or retrograde] hope . . . The one wing of the opposition, to which Schrödinger belongs', Pauli continues with marked irony, 'holds waves to be more beautiful than particles and therefore tries to eliminate the concept of a particle . . .'.[1] I cannot help feeling that the irony is misplaced: nobody wishes to 'eliminate' the concept of a particle, but some people felt, and still feel, that it may turn out not to be an ultimate and irreducible category of physics. Indeed, I should be surprised if Pauli or Born or any other member of the orthodoxy were to deny this. Why then attack Schrödinger for first opening up this problem, and for first opening our eyes to the possibility of unstable particles? Explaining a thing is not explaining it away.

To state briefly my own opinion with regard to the first question, I feel inclined to say that Schrödinger's theory according to which

[1] W. Pauli, *Dialectica* **8**, 1954, p. 115.

particles are fairly simple wave structures ('wave packets') is too simple, and is refuted by the character of particle transition. Particle structures are more complex, I think, and the wave packet should be interpreted as determining propensities (as explained below) rather than as a description of the particle's structure. But this or any other view of the matter must, if at all tenable, be heavily indebted to Schrödinger's truly revolutionary theory of particles. (*Cp.* section 26, below.)

The answer to the second question—whether the existence of particles is implied in the interpretation of the present quantum theory—of course depends upon the interpretation chosen. Born, Pauli and Heisenberg clearly imply that the existence of particles is part of the interpretation which they accept, which is the orthodox or Copenhagen interpretation, and that they differ in this point from Schrödinger. I find this a little surprising; for prior to this controversy over Schrödinger's article 'Are there Quantum Jumps?' I should have thought that the Copenhagen view on this matter was very different. I should have thought that, according to this view, particles and waves were *nothing but* two kinds of 'pictures' which we use in order to *apply* the quantum-theoretical formalism, and which are needed since every application must of necessity use classical models and ideas. The concepts of 'particles' and of 'waves' are used just because they are such classical ideas. As to their reality, or existence, I thought that the Copenhagen interpretation adopted the view that this question should not be asked, or at least that the answer would be that nothing like particles or waves exists, and that particles and waves are only classical concepts with which we describe certain aspects (complementary aspects) of whatever may exist. This, I thought, was the orthodox view. But you never can tell.

(This orthodoxy is even more elusive than orthodoxies usually are. There is no Bible available as yet for exegesis; thus exegesis must not be attempted by anybody outside the esoteric orthodoxy itself. And the orthodox exegesis is sometimes indistinguishable from a surreptitious change of front.)

As against this view, I should have thought that Schrödinger asserted with great force the real existence of atoms or particles, even though he tried to explain them as wave structures. I did not know

that, by saying that a table or a chair is made of timber, I should expose myself to the criticism that I have denied the existence of discrete tables and chairs, and that I have done so because I am 'in love' with timber (as Born says of Schrödinger's attitude towards waves[2]).

Nor can I recall that in the long history of the controversy concerning the wave theory of light, anybody ever said that those who maintained this theory were committed to the view that light—or more correctly, rays of light—did not exist. The case is perfectly analogous, since Schrödinger explained the transition from particle mechanics to wave mechanics as a step corresponding precisely to the transition from the ray theory of light (geometrical optics) to the wave theory.

But I shall not continue to investigate the attitude of the various interpreters of the existing quantum theory towards our problem 'Are there particles?'. Instead I shall indicate briefly what the answer would be from the point of view of the propensity interpretation.

The answer would be this. The present quantum theory is, viewed from the propensity interpretation, a particle theory. The waves merely determine the probable *states* of the particle, that is to say, the probability or the propensity of the particle's being in a certain place, or possessing a certain momentum; exactly as do the waves in the soldier's random walk. The waves do not constitute the particles, any more than the waves in the random walk example constitute the soldier. But just as the waves in the random walk example are characteristic of the conditions of the situation, or of the whole experimental set-up, so are the waves of the quantum theory.

I therefore declare myself unambiguously in favour of a particle interpretation, although I am in love with waves. I do not find any contradiction between being in love with an idea—believing that there is very much to be learned from it—and asserting, at the same time, that this idea does not fit a certain problem situation. Schrödinger's wave theory of the particle is full of promise, and it marks the beginning of a theory of particles. Yet at the same time, the present theory can be most easily interpreted as a particle theory.

To avoid any misunderstanding, I wish to make clear that I do not believe in a dualism of particles and waves or anything even faintly

[2]Born, *British Journal for the Philosophy of Science* 4, 1953, p. 98.

resembling it. As to the pilot waves of de Broglie, they can, I suggest, be best interpreted as waves of propensities.

15. *Position Space.*

The idea of a 'particle' has changed, we have seen, from that of something ultimate and indestructible to that of something changing and in need of explanation. Why then do we continue to use the term 'particle'? The question is not important, but also not wholly without interest, in view of the protests which Schrödinger's alleged disbelief in particles has aroused.

The strong inclination to speak of 'particles' (which I share) is doubtless connected with the intuitive idea that a particle is something located at any moment in a small region of space, whose connected track in space may, in principle, be followed through its various positions. By 'space' I do not mean here 'momentum space', or 'space-time', or any abstract space, but ordinary or, as we may say, *position space.*

A particle track in a Wilson cloud chamber, or in a Powell stack of photographic plates (which is fundamentally the same thing) is the kind of track in position space that I have in mind; thus Wilson chambers and Powell stacks are essentially parts of position space.

It is not without interest to point out that *all* our experiments with particles, and more generally with quantum mechanical systems, consist in interpreting measurements of positions. This assertion is even trivial, as a moment's thought will make clear. All measurements of velocities or momenta consist essentially in interpreting measurements of position. This is even true of runners, motor cars, or aeroplanes whose velocities we may be able to estimate by sight; if we wish to measure their velocities, we shall either measure two positions, or read off the position of a pointer on a speedometer. With atomic or subatomic particles, the situation is precisely the same. In measuring momentum, we measure position; and as a rule, more than one position is needed.

This is true of any method using spectroscopes, for it is the final position of the photon which is indicative of its frequency or momentum; and it is so even if we use a filter: the essential point is that the particle has *passed* the filter, which is a question of its location in position space.

Bohr has said something very close to the point I wish to make. For he writes '. . . it must be realized that . . . [the] use of space-time concepts in the description of atomic phenomena is confined to the recording of observations which refer to marks on a photographic plate, or to similar . . . amplification effects like the building of water drops around an ion in a cloud chamber'.[1] Bohr's remark contains a timely reminder that even the 'observation' of a part of the track of the particle in position space is, like all 'observations', actually an *interpretation* in the light of theories, in spite of its apparent intuitive obviousness. The measurement of the *momentum* of the particle will therefore be an interpretation of an interpretation; that is, an interpretation of higher order. (It may be said in passing that all our measurements are interpretations of a very high order; so high, indeed, that we often cannot even compare the interpretations as to whether they are of higher or lower order.)

The fact that all measurements of momentum go back to measurements of position is, as I shall explain in the next chapter, of considerable significance for an evaluation of Heisenberg's so-called 'Principle of Indeterminacy'.

[1] *Cf.* N. Bohr, 'Discussion with Einstein on Epistemological Problems in Atomic Physics', in *Albert Einstein, Philosopher-Scientist*, pp. 222 f.

TOWARDS A RESOLUTION OF THE PARADOXES OF QUANTUM THEORY

16. *Indeterminacy or Scatter?*

What I have said so far in this volume of the *Postscript* has been mainly introductory. There are two exceptions to this: my discussion of the objectivity of the quantum theory, and of the so-called 'reduction of the wave packet'. I now proceed to give an outline of the way in which the propensity interpretation resolves the various paradoxes of quantum theory. To this end I shall discuss, in the present section, the indeterminacy relations, and in the following two sections the experiment of Einstein, Podolsky, and Rosen, and the so-called 'two-slit experiment'.

My views with respect to Heisenberg's so-called 'indeterminacy relations' are practically unaffected by the transition from my old frequency interpretation of probability to the propensity interpretation. (This is due to the similarity of the two interpretations when it comes to explaining the reduction of the wave packet.)

The formula,

$$\Delta x\ \Delta p_x \geqslant h$$

explained in section 73 and discussed in section 75 of *L. Sc. D.*, I still consider to be part of the formalism which we have to interpret; and I still interpret it as a scatter relation. The only change is that I should now interpret the formula as a *singular* probability statement, and therefore as determining the propensity of a single particle to 'scatter': it predicts that the actual statistical scattering will be observed if we *repeat* the experiment in question many times, each time with a single particle.

144

Thus let Δx be the width of a slit in the (horizontal) x-direction and let a monochromatic ray of particles—photons or electrons—fall (vertically) upon the slit, as shown in Figure 4:

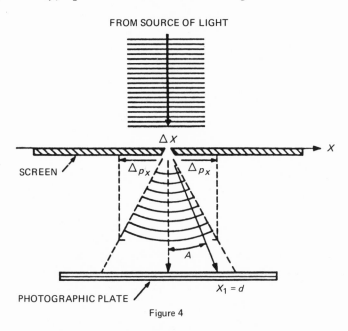

Figure 4

The formula tells us that the narrower we make the slit, the greater will be the propensity of the particles to scatter to the right or to the left from the vertical direction. *This is all.* The interpretation is objective, and nothing in it concerns the subject, its interference with the object, and its necessary nescience. The formula, even the interpreted formula, *cannot* contain anything about these matters, since the formulae of the calculus are interpreted (as in *L.Sc.D.,* section 75) as objective formulae.

But we can say a little more. Following what I said in *L.Sc.D.,* we can say that these scatter relations are *testable.* We test them by putting a photographic plate in the way of the ray as it emerges from the slit (see Figure 4). Our prediction is: if we repeat the experiment very often with a fixed width Δx of the slit, then we shall find that the particles blacken the plate in a region as indicated by the formula. And if we arrange for several sequences of the experiment, with varying width of the slit, then we shall find that the size of the

blackened region of the plate varies (roughly) inversely, in accordance with the scatter relations.

I do not think that anybody will deny that the scatter relations *can* be interpreted and tested in this way: the experiment can easily be carried out.[1]

But I now assert—as I did before—that this test implies (i) that we must have retrodictive *knowledge* beyond that permitted by the usual subjective interpretation, and (ii) that this knowledge is not 'meaningless' or 'redundant' or 'gratuitous', but highly relevant; for it is precisely this retrodictive knowledge which allows us to *test* the theory.

(i) The blackening of our photoplate is the result of many single experiments. Let us take one of these: it will result in one single hit of some particle, say, at the place x_1 of the plate. This allows us to determine the angle A (between the vertical direction and the deflected direction taken by the particle), and therefore the momentum of the particle in the x-direction. Thus our measurement x_1 in position space may in this case also be interpreted as a measurement of the momentum. As a result, we obtain position and momentum of the particle with a precision far exceeding the indeterminacy relations.

(ii) Admittedly, the knowledge so obtained is retrospective or retrodictive knowledge, as far as the momentum is concerned. (As far as the position is concerned, the knowledge is predictive, in that we should obtain on a second plate, put underneath the first, a spot very close to the first—provided our particle was not absorbed by the first plate.) For the particle will have been slowed down by the interaction with the plate.

(iii) As far as the retrodictive information is concerned, I assert that we *must* accept the view that the track of the particle can be retraced right back to the slit. It will be remembered (see *L.Sc.D.*, section 73, text to notes 5 and 6) that Heisenberg suggests that this calculation of the past track of the particle has a dubious status, and that it is a 'matter of taste' whether one attaches any physical

[1][See further discussions of this issue by Popper in *Objective Knowledge*, 1972, pp. 301–4; and 'The Propensity Interpretation of the Calculus of Probability, and the Quantum Theory', *op. cit.* See also the more elaborate discussion in the 'Introduction' to the present volume, sixth thesis. Ed.]

significance to it or not. (Schlick and others declared the retrodictive knowledge to be 'meaningless'.)

I am of a different opinion. The scatter relations predict, we have seen, that the scatter of the momentum increases with the contraction of the slit. *We must therefore test the scatter immediately after* (in our figure, below) *the slit.* This we can do only by tracing the track of the particle back to the slit. This shows that the retrodictive calculations which the formalism permits are not redundant but necessary for testing the assertion made about the scatter relations.

The situation is, as usual, very much like coin tossing. The theory of a tossing machine may tell us that the propensities to fall head or tail are equal upon the coin being tossed by the machine. In order to test this assertion, we have to observe the results of various tosses. These observations give us retrodictive knowledge only. They do not give rise to prediction about the further adventures of the coin—except for the somewhat trivial prediction that if I look at the same coin again, I shall find that it still lies heads up—exactly as a second photographic plate or film underneath the first will confirm the position measurement of the particle carried out by the first.[2]

17. *The Experiment of Einstein, Podolsky, and Rosen.*

We must then attribute a physical meaning and a relevance to the retrodictive calculations which allow us to determine the position *and* momentum of the particle. This has the following clear consequence: the particle always *has* position and momentum; and the theory does not predict them sharply simply because it is a *probabilistic theory* which only informs us of the propensities to take up various positions if the momentum is given, and of the propensity to take up various momenta if the position is given. (If you select the blue-eyed, or the brown-eyed, etc., you will get various distributions of the colour of hair. Similarly, if you select according to the colour of the hair, you will get various distributions of the colour of the eyes. Can you select according to both? Of course, but then there is no room left for any *predictive* statistical or propensity theory relating one property to the other. The whole quantum mystery is hardly more mysterious than this.)

[2]Compare the discussion of Heitler in sections 7–9 above.

From this point of view, there is no doubt that the particle *has* position and momentum, and that we can even know it, although only retrospectively, because the theory—which is a propensity theory like that of coin tossing—is only concerned with statistical or propensity predictions.

In *L.Sc.D.*, section 77, in order to show that the particles actually do have position and momentum, I introduced an experiment. This particular experiment turned out to be untenable.

Shortly afterwards, however, Einstein, Podolsky, and Rosen published an experiment which also purported to show that we must attribute to the particles both positions and momenta, *according to the theory*—in strongest contrast to the Copenhagen interpretation.[1]

This thought-experiment (of which Einstein kindly informed me in a letter)[2] was at once severely criticized by Bohr.[3] But Bohr's reply is, in my opinion, invalid. This also remained Einstein's opinion to the end.

The idea of Einstein, Podolsky, and Rosen is very simple. We may let two particles *A* and *B* collide. *A* flies away. *B* can be observed and measured. We may choose arbitraily whether to measure the position of *B* or its momentum. If we decide to measure the position of *B* we can then calculate the position of the far away *A*. If we decide to measure the momentum of *B*, we can then calculate the momentum of the far away *A*.

Thus *A* must have both position and momentum, even though the theory does not enable us to calculate both predictively at the same time. This beautiful and simple argument seems to me conclusive.

It also has far-reaching consequences. According to the orthodox view, the quantum mechanical particle is distinguished from more ordinary or 'classical' particles precisely by the fact that it is meaningless to attribute to it both position and momentum at the same time. We must not seriously try to visualize a particle of this

[1]*Cp.* A. Einstein, B. Podolsky, and N. Rosen, *Physical Review*, ser. 2, **47**, 1935, pp. 777 *ff.* [See further discussion by Popper of the Einstein-Podolsky-Rosen experiment in 'Particle Annihilation and the Argument of Einstein, Podolsky, and Rosen', *op. cit.* This article contains some points and arguments which Popper no longer upholds. But see section I of the 'Preface 1982' to the present volume. Ed.]

[2]This letter contains a concise and, in my opinion, an improved version of the argument. It is printed in *L.Sc.D.*, Appendix *xii.

[3]N. Bohr, *Physical Review*, ser. 2, **48**, 1935, pp. 696 *ff.*

kind; but the particle behaves as if its momentum would become 'smeared' over a considerable range, the moment it 'takes up' a definite position; and *vice versa*, as if the position would become 'smeared' when it 'takes up' a definite momentum. The 'taking up' is a kind of reaction, stimulated by *interfering* with the particle, that is to say, by subjecting it to a measuring experiment.

This interpretation of the formalism appears to me refuted by the ideas of Einstein, Podolsky, and Rosen. *For there is no interference with the particle A* which indeed may be far away at the moment we interfere with *B* by measuring either its position or momentum.

Although this clear and simple argument is decisive, it has not been accepted as such. I believe that the reason for its rejection is simply Bohr's authority, not counter-argument. For in contrast to Einstein's simple and clear ideas, Bohr's counter-argument is very hard to understand. Einstein never understood it (as he told me), and I have not met any physicist who could present it in anything like a clear and simple form. I do think that I have managed to understand the gist of it, but I am not at all certain.

As I understand it, Bohr's counter-argument is this. When we make up our minds (after *A* has departed) whether we shall measure the position of *B* or its momentum, then we commit ourselves thereby to choosing between two different experimental arrangements; to put it simply, we have to choose between an experimental framework which would allow us to relate *B* to a *position space*, and another experimental framework which would allow us to relate *B* to a *momentum space*. The two experimental arrangements amount to setting up two different co-ordinate systems. The arrangements are exclusive of each other (or 'complementary'), the argument continues, and so we cannot combine the two spaces or co-ordinate systems into one. (Bohr dwells on this exclusiveness or 'complementarity', strangely enough, although it clearly was not an issue here since his opponents *accepted* the assumption that we must *choose* whether we measure the position or the momentum of *B*, and that we cannot measure both at a time.) As a result, we find that if we choose the positional framework, *A* becomes (indirectly, by way of *B*) related to this framework; and if we choose the momentum framework, *A* becomes related to this other framework. The conclusion is that *A* cannot be related to both co-ordinate systems at the same time, so that it becomes meaningless (if I understand it

properly) to attribute to *A* a momentum if it is unrelated to the momentum framework, or a position if it is unrelated to the position framework.

This is the best I can make of Bohr's argument. I do not at all assert that this is what he wanted to say. My criticism, which follows here, is therefore not directed against Bohr, but only against my own version of his reply to Einstein, Podolsky, and Rosen.

My criticism consists of two parts. First, the reply to Einstein and his collaborators consists of a surreptitious change of the theory which Einstein attacked, of a shifting of the ground. Before the shift, we heard that the particle reacted to measurements by taking up a sharp value of one variable and becoming smeared in another. But now we hear that what was meant was something completely different and much more harmless. There was nothing more in it than that, at times, one co-ordinate system is applicable, and at other times another co-ordinate system, but never both together. This leaves completely open what the particle itself does. The suggestion that it is *meaningless* to speak of position and momentum unless we can relate the particle to the corresponding co-ordinate system is as gratuitous as the assertion that it is meaningless to call a man wealthy unless he pays income tax.

The second part of my criticism is that the reply to Einstein and his collaborators misses the point. The main argument was neither that the indeterminacy formulae are invalid, nor that the quantum theory is inconsistent (nor even that it is incomplete, although Einstein said so). The main argument was the simple point that if I can freely decide—and if I can successfully carry out my decision *without interfering with you*—whether I wish to measure your size or your weight, then it is unreasonable to say that you do not have both size and weight at the same time but that you adopt the one or other by adjusting yourself to the stimulus of the measurement.

Reasonableness was the point at issue. The question is not whether by a subtle and highly scholastic argument we may continue to uphold an untenable position. The question is whether we should think critically and rationally in physics, or defensively and apologetically. (One can always escape refutations by a conventionalist stratagem or by similar methods. *Cf. L.Sc.D.*, sections 19*f.*)

Before leaving the experiment of Einstein, Podolsky, and Rosen,

I wish to indicate why I feel doubtful whether the so-called pilot wave theory of de Broglie, rediscovered by David Bohm, is unacceptable. I feel so because of Bohm's reply to Einstein, Podolsky, and Rosen.[4] Bohm does accept the view that particles have both sharp positions and momenta. But he interprets the theory and especially the indeterminacy relations in such a way that he cannot accept the simple conclusion we have drawn here. Instead he asserts that by measuring the position of B, we do disturb A—with superluminal velocity. Thus if we measure the position of B, we do interfere with B, and also with A, however far it may have moved away. I agree with Einstein that this suggestion has nothing to recommend it. And I should say the same about a later reply of Bohm's to Einstein.[5] Bohm's last remark in this paper deserves to be quoted. He says that he 'would admit only two valid reasons for discarding a theory that explains a wide range of phenomena. One is that the theory is not internally consistent, and the second is that it disagrees with experiment'. Against this, it must be pointed out, first, that Einstein never suggested discarding the quantum theory; he only discussed the acceptability of various competing interpretations. Secondly, there may be valid reasons for criticizing a theory other than those stated by Bohm. (A theory may not be clearly refuted; but may fail here or there as in the explanation of the electronic charge or of particle transition; or it may perhaps explain the phenomena less satisfactorily than another one.)

18. *The Two-Slit Experiment.*

I wish to discuss here a type of experiment due to Young which, I believe, is at the very root of the whole trouble. It has been discussed again and again by Bohr, and has no doubt deeply impressed him. Indeed, I believe that it cannot be understood without difficulty except perhaps in terms of the propensity interpretation.[1]

[4]Contained in D. Bohm, *Physical Review* **85**, 1951, pp. 166 *ff.* and 180 *ff.*

[5]*Cp.* D. Bohm in *Scientific Papers Presented to Max Born*, 1953. The quotation is from p. 18. [But see now Popper's new discussion in 'Preface 1982' to this volume. Ed.]

[1]I discussed the experiment in Appendix v ot *L.Sc.D.*, but I could deal with it there only in a 'statistical re-interpretation'—which, however, is not without interest.

The arrangement of the experiment consists of a source of light (or of electrons); of a first screen with a narrow slit, to ensure that our source is as small as possible; of a screen with two slits which let the light through, and of another screen (or photographic plate) on which we may observe interference fringes, according to the wave theory of light. (*Cp.* figure 5.)

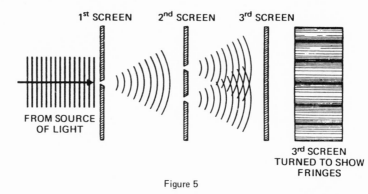

1st SCREEN 2nd SCREEN 3rd SCREEN

FROM SOURCE
OF LIGHT

3rd SCREEN
TURNED TO SHOW
FRINGES

Figure 5

The main problem is this. If we dim our source of light until only one quantum (one particle) is emitted in one experiment, and repeat this experiment once a day, and superimpose the results of many repetitions, we still get interference fringes (or so we assume). These fringes change if one of the two slits is closed: if we open only the upper slit for a time, and then only the lower for a time, the result obtained will be quite different from that obtained if both slits are open at the same time. Thus the two slits *co-operate* in producing the fringes, or in determining the probability that the particle will reach one point on the second screen rather than another. (This co-operation can be understood with the help of Huygens's principle.) But each particle can only pass through *one* of the slits. How can it be influenced by the fact that the other slit is open (or closed)?

From the point of view of the propensity interpretation, the reply to this question is simple: *it is the whole experimental arrangement which determines the propensities.* The possible results of any one experiment are clearly different in the case of both slits being open from the case where only one is open. But propensities are, as we know, dependent upon possibilities. We can therefore well understand the fact that the results differ. The difference follows from the

152

mathematical theory—from the wave character of the equations which determine the propensities.

Thus the particle will pass through only one of the slits, and in a certain sense will remain uninfluenced by the other slit. What the other slit influences are the propensities of the particle relative to the entire experimental arrangement, not the particle itself: the propensities for reaching the one point or the other on the second screen.

The situation is similar to the one in a well-known game of chance (pin board, or bagatelle table): little balls roll down a board hitting pins and moving on either to the right or to the left at each pin. If one ball rolls down, there will be many pins which it never touches. They do not influence the ball. But they influence the propensities inherent in the experimental arrangement: the propensities would change if these pins were shifted or removed. And upon frequent repetition of the experiment, the statistical results would change with the changed propensities.[2]

This is our fundamental answer. To understand it fully, we take any other arrangement which might enable us to find out through which of the two slits the particle passes. (For example, a beam of light which would allow us to 'see' or photograph an electron when it emerges from, or enters, a slit.) Any such arrangement will change the experiment; and calculating the experiment will show us that the change does away with the fringes. (This we know because the theory does entail the Heisenberg relations.) We interpret this to mean that the propensities—which depend on the arrangement— have changed. We do not even need to ask whether the change is due to an interference with the electron (as in the case of the light beam guarding the slits) or only interfering with the possibilities, that is, the propensities themselves (as in the case of shutting one slit): all we need to know for all these cases is that the wave equation which allows us to determine the propensities entails the Heisenberg scatter relations, and that these limit the possible predictions.

[2][See Popper's elaboration of this discussion in 'Quantum Mechanics without "the Observer" ', *op. cit.*, pp. 33–6, reprinted as the Introduction to the present volume. I know of only one attempt to rebut Popper's analysis here: Paul Feyerabend's 'On a Recent Critique of Complementarity', in *Philosophy of Science* **35**, pp. 309–31; and **36**, pp. 82–105. Feyerabend's criticism is shown to be without merit in my 'Critical Study: The Philosophy of Karl Popper, Part II: Consciousness and Physics', *Philosophia* **7**, July 1978, pp. 694–5. See also my discussion in the Introduction above, footnote 65. Ed.]

Thus I see no difficulty here, nor any problem which is specifically connected with quantum theory—that is to say, with Planck's constant h. We are faced, rather, with a problem which reappears in various probabilistic theories, for example, in diffusion theory.

The following difficulty may be raised in connection with the two-slit experiment. Consider a photon coming from far away and approaching one of the two slits. Will its propensities be affected if, while it is on its way, the other slit is shut or opened?

The correct reply to this question is that it was wrongly put because there is no possible experimental set-up which corresponds to it. For the only ways of controlling the approach of the photon towards one of the two slits would amount to a rearrangement of the experiment fundamentally equivalent to closing the other slit. Thus the question cannot be answered, at least not within our theory.[3]

It is not without interest to elucidate this reply with the help of our pin table (bagatelle table) set-up. Here a certain pin P may be removed without affecting the propensities, once the ball has passed the last place from which P may be reached. This might suggest the following reply to the question we have just dismissed: once the photon has proceeded to a point from which light can no longer reach the second slit, opening or shutting it will no longer influence the propensities. But any such reply would be mistaken. In the case of the pin table the experimental arrangement—and thus the propensities—may conceivably not be affected in any significant way if we watch closely and remove the pin P once we have seen the ball passing the last place from which P may be reached. Nevertheless, we must realize that even in this case, there are two entirely different experimental set-ups: the one which provides for the removal of P and the other which does not so provide. That these two different experiments may happen to determine equivalent propensities is perfectly possible, but whether or not they do so in fact will depend upon the details of the arrangements. (For example, the removal of the pin may generate a vibration in the table which may influence the propensities.) In the case of the two-slit experi-

[3][*(Added 1981) This is another place where my wording may be misunderstood to imply that propensities are relative to *experimental* set-ups only, rather than to the physical situation. See n. 63 to the Introduction to the present volume.]

ment, the theory tells us that *any* arrangement allowing us to ascertain that the particle (photon, electron, etc.) has passed a certain place will unavoidably change the experimental set-up—in such a way that it changes the propensities with it.

These considerations have some bearing on Bohr's famous principle of complementarity. What is behind this principle is, I am afraid, not much more than this: that different experimental arrangements are always different; and that any two experimental arrangements, however similar, always exclude one another, in the sense that they can never be combined. For any attempted combination will create a new experimental set-up. How the propensities are affected by such a change is a matter for the theory to decide.

It is misleading to suggest that experiments designed to measure position are exclusive of (or 'complementary' to) experiments designed to measure momentum, while two experiments designed to measure position are not exclusive of each other and always produce the same result. The truth is that two experiments to measure position can also not be combined without producing a third experiment—an experiment in which position is measured twice in succession. And what the theory tells us is that (since the particle 'has' a position) these two experiments, if carried out correctly, will yield compatible results. Similarly the view that measurements of position and momentum cannot be combined in the same experiment is also mistaken: for as we have seen, all so-called measurements of momentum are in fact high-level theoretical interpretations of what is, upon a somewhat lower level of interpretation, a measurement of position. (*Cp*. section 15.)

This becomes clear if we consider Bohr's own arrangement for measuring momentum by spring balances. (*Cf*. Schilpp volume on Einstein, *op.cit.*, pp. 220, 227.) Whenever we *use* the spring balance, we have to read two *positions* of the pointer (or the difference between two positions).

A considerable number of experiments have to be treated in the same way as the two-slit experiment. A famous one is that of a semi-reflecting mirror which is placed in the way of a photon. The two divided 'beams' may be reflected again, and may be made to interfere, producing fringes. Thus, one could argue, the photon must have been in both half-beams; but if we intercept the two

half-beams by photographic plates, we shall find (according to the theory) that the photon is always undivided, and that it is either in the one beam or in the other.

The explanation of this apparent contradiction is, of course, that the *photon* is always undivided in one of the half-beams, and that the *propensities* can be calculated by the theory which tells us that they behave like wave trains (or 'beams') which may be split up, and which may then interfere.

Dirac's famous example of polarization by a tourmaline also belongs to this category of experiments. There is no reason whatever to assume (as Dirac suggests) that the single photon is in the two different and incompatible states at one time; any more than to assume that it moves through both slits at one time. There are different possibilities, different virtual states, inherent in the experiment; and these different possibilities are weighed or measured by the propensities.

19. *An Apology for Having Been Controversial.*

I wish to explain in this concluding section my motives for criticizing some physicists for whom I have the greatest admiration.

A very serious situation has arisen. The general anti-rationalist atmosphere which has become a major menace of our time, and which to combat is the duty of every thinker who cares for the traditions of our civilization, has led to a most serious deterioration of the standards of scientific discussion. It is all connected with the difficulties of the theory—or rather, not so much with the difficulties of the theory itself as with the difficulties of the new techniques which threaten to engulf the theory. It started with the brilliant young physicists who gloried in their mastery of the tools and look down upon us amateurs who had to struggle to understand what they were doing and saying. It became a menace when this attitude hardened into a kind of professional etiquette.

But the greatest among contemporary physicists never adopted any such attitude. This holds for Einstein and Schrödinger, and also for Bohr. They never gloried in their formalism, but always remained seekers, only too conscious of the vastness of their ignorance.

Why has Schrödinger been answered only by arguments which

cannot for a moment be taken seriously? I believe the reason is that his distinguished critics simply do not take argument seriously any longer. Had Schrödinger surprised them again with a new formalism, they would again have listened to him most attentively. But *mere words* no longer interest our specialists; not even if they come from a man who has done at least as much for their subject as any other. If a great physicist like Schrödinger is treated in this fashion—and Einstein was treated just as badly—what can a mere amateur like myself expect if he dares to differ from the professionals?

It is partly in an attempt to clarify the situation, and partly in an attempt to show that we must restore ordinary rational discussion to its rightful place, that I have gone into the details of the controversy opened by Schrödinger.[1]

I admire everybody who has the power to help in the great task of discovery. But I am not impressed by the specialist—the expert—as such. We must not allow ourselves to be bowled over by imposing formulae: that is the road to perdition. Admiration for the impenetrable expert puts a premium on sheer difficulty of expression and tempts us to mistake obscurity for profundity. We must insist upon clear thinking, simplicity, and intellectual responsibility in discussing subjects which, after all, ought to be everybody's concern.

When I recently looked into the controversy centered round Boltzmann,[2] I came to the conclusion that our standards of rational discussion have seriously deteriorated since. The decline started with the First World War and with the growth of the technological and instrumental attitude towards science. But it was much aggravated by the increased prestige of sheer difficulty as such. It rests with us to put a halt to this development by reviving the traditional values of understandability and simplicity.

Having said so much about the motive behind my controversial tone, I wish to make clear that there was also another. This was my boundless admiration for the work of Einstein and of Schrödinger.

[1] I do not wish to extol Schrödinger as an authority beyond criticism. On the contrary, I feel highly critical of much that he has written: of his determinism, his attempts to argue in favour of a subjective interpretation of probability, and especially his theory of the Arrow of Time, and its connection with the Second Law of Thermodynamics. [See Popper's discussion of Schrödinger in *Unended Quest*, 1976, section 30. Ed.]

[2] [See Popper's discussion of Boltzmann in *Unended Quest*, 1976, section 35. Ed.]

Not that I think that criticizing Einstein or Schrödinger is *lèse majesté*; after all, I have myself criticized Einstein's determinism, his and Schrödinger's intermittent inclination towards a subjective theory of probability, and in part also Schrödinger's particle theory. But I feel that the manner in which they have sometimes been criticized is symptomatic of lack of appreciation.

About Einstein I hardly need to say what anybody who has any knowledge of the history of human thought can fail to realize: that perhaps no other man since Galileo, Kepler, and Newton has done so much to open our minds.

As to Schrödinger, many of the physicists of the new generation know him mainly from the text-books, as the famous author of the wave equation. This is a pity. For Schrödinger's *Collected Papers on Wave Mechanics* are a classic. I think they are unique. The directness of Schrödinger's approach, the depth and the sheer beauty of his ideas, his lucid presentation, his wonderful detachment and self-irony, the fact that he looks for, and finds, a refutation: the whole is a wonderful human document, a work of art, an adventure of ideas which has few parallels.

And it is *physics*, in the grand style. Schrödinger is not only the real father of the *formalism* of the quantum theory, he is first of all a physicist who tries to understand the physical world in which we live. He is, like Einstein, a true heir to the speculations of Faraday, and he is the first to have shown us that matter may one day be explained as a disturbance of something that is not in its turn material.

I do not think there is any chance that this something will turn out to be 'mind-like' or 'spiritual', as some apologists of the spiritual try to make out. Nor do I think that the human mind or the human spirit needs an apologist; or that we can anticipate its future adventures and discoveries.

As to the great founders of the new theory of matter—from Planck, Einstein, Bohr and de Broglie to Schrödinger, Pauli, Born, Heisenberg and Dirac—I may be allowed to apply to them the words of a master. 'I cannot express strongly enough my unbounded admiration for the greatness of mind of these men', Galileo wrote of the founders of the heliocentric system, 'who conceived it, and held it to be true . . . in violent opposition to the evidence of their own senses.'

158

A METAPHYSICAL EPILOGUE

There is something besides extension, even something that is prior
to extension.

<div align="right">LEIBNIZ[1]</div>

INDETERMINISM and the propensity interpretation of probability
allow us to paint a new picture of the physical world. According to
this picture, of which only the roughest sketch can be given here, all
properties of the physical world are dispositional, and the real state
of a physical system, at any moment, may be conceived as the sum
total of its dispositions—or its potentialities, or possibilities, or
propensities.

Change, according to this picture, consists in the realization or
actualization of some of these potentialities. These realizations in
their turn consist again of dispositions or potentialities—
potentialities, though, that differ from those whose realizations
they are. This view corresponds closely to the commonsense view of
the world. When a student enters the university there is a certain
possibility inherent in this situation of his sitting his finals; of his
answering the examination questions more or less successfully; and
of his obtaining a degree. When he sits his finals he is realizing the
first of these possibilities or potentialities; and at the same time, the
new situation which has arisen changes the other two possibilities.
In fact, the sitting of his finals can be said to be essentially a
possibility of his answering examination questions more or less
successfully, and of obtaining his degree. When he has given his
answers, and when he has thus realized another set of possibilities or

[1]'Est aliquid praeter extensionem imo extensione prius', quoted by Kant in
*Gedanken von der wahren Schätzung der lebendigen Kräfte und Beurteilung der
Beweise*, 1746, end of the first main section, § 1, in *Werke*, ed. Ernst Cassirer, Vol.
I, p. 15.

potentialities, a new situation has been created which again changes the potentialities of his obtaining a degree. When he ultimately gets his degree, and thus again realizes certain potentialities, then the new situation creates new potentialities—relating, say, to his making use of his degree, or not making use of it, in his career. We thus obtain a picture of the world which is at once dualistic and monistic. It is dualistic in that the potentialities are potentialities only relative to their possible realizations or actualizations; and it is monistic in that the realizations or actualizations not only determine potentialities, but may even be said to be potentialities themselves. (But we should perhaps avoid saying that they are 'nothing but' potentialities.) Thus we may describe the physical world as consisting of *changing propensities for change*. Although these propensities do not in general determine future changes, they may determine, in some fields of physics at least, the probability distributions—which may include probabilities equal to 1—of the various possible future changes.

This approach contains the suggestion of a theory of matter, or of particles, according to which these are interpreted as realizations of potentialities or propensities, and at the same time as consisting of potentialities or propensities in their turn.

One of the main points of this approach is the suggestion that it may be possible, in this way, to give an indeterministic re-interpretation of Einstein's deterministic programme, and at the same time an objectivistic and realistic re-interpretation of quantum theory. The aim is a picture of a world in which there is room for biological phenomena, for human freedom, and for human reason.

20. *Metaphysical Ideas and Research Programmes, and the History of Physics.*

One day somebody should write the history of physics as a *history of its problem situations*. (This, incidentally, is the way in which all history should be written, I believe, including political history.) Problem situations as they affect the history of physics (as opposed to, say, political history) may be analyzed almost completely in purely logical terms, provided we take notice of the metaphysical ideas which contribute to the creation of problems and

which largely determine the direction in which we seek for a solution.[1]

In science, problem situations are the result, as a rule, of three factors. One is the discovery of an inconsistency within the ruling theory. A second is the discovery of an inconsistency between theory and experiment—the experimental falsification of the theory. The third, and perhaps the most important one, is the relation between the theory and what may be called the *'metaphysical research programme'*.[2]

In using this term I wish to draw attention to the fact that in almost every phase of the development of science we are under the sway of metaphysical—that is, untestable—ideas; ideas which not only determine what problems of explanation we shall choose to attack, but also what kinds of answers we shall consider as fitting or satisfactory or acceptable, and as improvements of, or advances on, earlier answers.

By raising the problems of explanation which the theory is designed to solve, the metaphysical research programme makes it possible to judge the success of the theory as an explanation. On the other hand, the critical discussion of the theory and of its results may lead to a change in the research programme (usually an unconscious change, as the programme is often held unconsciously, and taken for granted), or to its replacement by another programme. These programmes are only occasionally discussed as such: more often, they are implicit in the theories and in the attitudes and judgements of the scientists.

I call these research programmes 'metaphysical' also because they result from general views of the structure of the world and, at the same time, from general views of the problem situation in physical cosmology. I call them 'research programmes' because they incorporate, together with a view of what the most pressing problems are, a general idea of what a satisfactory solution of these problems would look like (*cp. Realism and the Aim of Science*, Volume I of this *Postscript*, section 15). They may be described as *speculative*

[1][See Popper on problem situations and historical writing in *Conjectures and Refutations*, 1963, pp. 66–96; *The Self and Its Brain*, 1977, pp. 151 *ff; Objective Knowledge*, 1972, pp. 166–190. Ed.]

[2][Cp. Popper's discussion of metaphysical research programmes in *Unended Quest*, section 33; and in the 'Preface 1982' to the present volume. Ed.]

physics, or perhaps as speculative anticipations of testable physical theories.

In the present section, I shall first give, in approximately historical order, a bare list of ten of the more important metaphysical research programmes that have influenced the development of physics since the days of Pythagoras and Heraclitus; and more especially, since Parmenides, when the *problem of change* became the fundamental problem of metaphysics and of physics.[3]

After having given this list, I shall attempt to add, as an illustration, an outline of the sequence of problem situations that led from Descartes's programme towards Faraday's.

1. *Parmenides's Block Universe.* The nothing (void, empty space) cannot exist: the world is full, it is one block. Motion and change are impossible. A true picture of the world must be rational; that is, based on deduction, and upon the principle of non-contradiction.[4]

2. *Atomism.* Motion, and therefore change, are real. Thus the world cannot be full; the void must exist. The world consists of atoms *and* the void—of 'the full' *and* 'the empty'. *All change is explicable by the movement of the atoms in the void.* There is no qualitative change—only movement and structural change, that is, rearrangement. The void is the space for the possible movements and positions of the atoms.

3. *Geometrization.* An early Pythagorean programme had been the arithmetization of cosmology (including geometry); it broke down with the discovery of the irrationals. Plato turned the tables on this programme; he first conceived the geometrization of cosmology (including arithmetic). The physical world is space filled by matter. Matter is *formed* (or shaped, or moulded) space, and since geometry is the theory of shapes, and of space, the fundamental properties of matter are explained geometrically (in the *Timaeus*). The geometrization of the cosmos and of arithmetic is carried on by Eudoxus, Callippus, and by Euclid. (Euclid did not

[3]*Cp.* my paper 'The Nature of Philosophical Problems', in *Conjectures and Refutations*, 1963, pp. 66–96.

[4][See also Popper's discussions of Parmenides in *Conjectures and Refutations*, 1963, pp. 38, 79n, 80–1, 142, 145 *f*, 147, 150, 159, 400–1, and 405–13; and in *The Self and Its Brain*, 1977, pp. 5, 15, 153. Ed.]

intend to write a textbook of geometry, but to solve the problem of a geometrical theory of the irrationals, and other fundamental problems of Platonic cosmology.[5])

4. *Essentialism and Potentialism.* For Aristotle, space (*topos*, position space) is matter, and pure geometry loses its central place which is taken up by a dualism of matter and form (or essence). *The form or essence of a thing inheres in it, and contains its potentialities.* These realize themselves for the sake of their *final* cause, the end, the aim. (The good is self-realization.)[6]

5. *Renaissance* physics (Copernicus, Bruno, Kepler, Galileo, Descartes) is largely a revival of Plato's geometrical cosmology, his *antecedent* causes (*'universale ante rem'*), and his hypothetico-deductive method. (A little later there is also a revival of atomism.) It turns into:

6. *The Clockwork Theory of the World* (Hobbes, Descartes, Boyle). The essence or form of matter is identical with its spatial extension. (This is a combination of Platonic and Aristotelian ideas.)[7] Thus all physical theories must be geometrical. All physical causation is push or, more generally, action at vanishing distance. All qualitative changes are quantitative-geometrical movements of matter; for example, of the fluid of heat (caloric), or of magnetism, or of electricity. (Compare also the Bernoullian conjecture that atoms are vortices of the ether.)

7. *Dynamism.* All physical causation is to be explained either by push, or else by central attractive forces (Newton). Every *change* of a physical state is functionally dependent upon another *change*. (Principle of differential equations.) According to Leibniz, push must also be explained by forces—by central repulsive forces—because push can only be explained if matter is space *filled by (repulsive) forces.* Thus Leibniz has a dynamical and structural theory of matter. (The theory of central forces was further developed by Kant and by Boscovich.)[8]

[5]See my paper 'The Nature of Philosophical Problems', *op. cit.* [See also 'Back to the Pre-Socratics', in *Conjectures and Refutations*, 1963, pp. 136–165. Ed.]

[6][See Popper's discussion of Aristotle's essentialism in *The Open Society and Its Enemies*, 1945, Chapter 11. Ed.]

[7][See further discussions by Popper in 'Philosophy and Physics', *Atti del XII Congresso Internazionale di Filosofia* (Venice, 1958) 2, Florence, 1960, pp. 367–74; and in *The Self and Its Brain*, 1977, especially Chapters P1 and P5. Ed.]

[8][See further discussions by Popper in *The Self and Its Brain*, 1977, pp. 6, 177–96, Ed.]

8. *Fields of Forces* (Faraday, Maxwell). Not all forces are central forces. There are changing fields of (vectorial) forces whose local changes are dependent upon local changes at vanishing distances. (Principle of partial differential equations, combining Newton's and Descartes's principles of causality.) Matter—that is, atoms or molecules—may become explicable in terms of fields of forces, or of *disturbances* of fields of forces. (*Cp.* Bernoulli's theory mentioned above, under point 6.)

9. *Unified Field Theory* (Riemann, Einstein, Schrödinger). Geometrization of the fields, gravitational as well as electromagnetic. Maxwell's field theory of light is generalized to a field theory of particles, and thus of matter. Matter is predicted to be destructible (with the confirmation of this prediction, the clockwork cosmology—that is to say materialism—is refuted) and to be inter-convertible with radiation, that is to say, with field energy, and thus with geometrical properties of space. However, the view that matter is a disturbance (vibration) of the field is opposed by:

10. *The Statistical Interpretation of Quantum Theory* (Born). Since Einstein's photon theory it becomes doubtful whether even light is nothing but a Maxwellian disturbance, a vibration of the field. For with every train of vibrations (light waves) there is associated a particle-like entity, a photon, which is emitted by one atom and absorbed by one atom. According to de Broglie, there is a similar dualism of particle and wave for particles of matter. Now this dualism is interpreted by Born in a way which may be described almost as a reversion to atomism in its original sense: what exists are corpuscles or particles; and the field and its vibration merely represent the mathematical instruments of an indeterminist particle-physics by which we may calculate the purely statistical probability of finding a particle in a certain state. This view (which I largely subscribed to in *L.Sc.D.*, 1934) appears to be incompatible with the programme of a monistic unified field theory of matter; and what looks like the majority of physicists take the success of the statistical quantum theory to indicate that the Faraday-Einstein-Schrödinger programme of a field theory of matter has to be given up.[9]

[9]Heisenberg expressed this view very clearly in 'Der Begriff "Abgeschlossene Theorie" in der modernen Naturwissenschaft', *Dialectica* 2, 1948, pp. 334–6. I

This brief survey may help us to understand the fundamental problems of physical cosmology, and why they were fundamental. I have in mind such problems as the problem of change in general; of matter and space (of atoms and the void); of the spatial structure of the universe; of causation (of action at a distance or at vanishing distances; of forces; and fields of forces); of the (atomic) structure of matter, and especially of its stability, and the limits to its stability; and of the interaction of matter and light.

It is interesting to note that there have been only three theories of change so far: atomism, which explains qualitative changes by quantitative movements of matter; Aristotle's theory of potentialities and their actualization or realization, which is a qualitative theory; and the theory of disturbances (vibrations, waves) of fields—like atomism a theory which aims at explaining qualitative changes quantitatively, but by changing intensities rather than by the movement of extended matter. [*(Added 1981) Heraclitus, I think, proposed only a programme; and Parmenides's theory was one of non-change.]

Such research programmes are, generally speaking, indispensable for science, although their character is that of metaphysical or speculative physics rather than of scientific physics. Originally they were all metaphysical, in nearly every sense of the word (although some of them became scientific in time); they were vast generalizations, based upon various intuitive ideas, most of which now strike us as mistaken. They were unifying pictures of the world—the real world. They were highly speculative; and they were, originally, non-testable. Indeed they may all be said to have been more of the nature of myths, or of dreams, than of science. But they helped to give science its problems, its purposes, and its inspiration.

An illustration, if only a sketchy one, may be useful here, and I will therefore conclude this section by giving an outline of the history of a particular problem situation—that of the problem of matter since Descartes. (It is part of the history of the transition from the clockwork theory of the world to dynamism, that is to say, from programme 6 to 7.)

criticized his view briefly in my 'Three Views Concerning Human Knowledge', 1955, reprinted in *Conjectures and Refutations*, 1963, pp. 113 f. Heisenberg later changed his mind. [See the preceding chapters of this volume of the *Postscript*. Ed.]

The history of the problem of matter has been sketched before, notably by Maxwell (in his wonderful article 'Atom' in the 9th edition, 1875, of the *Encyclopaedia Britannica*). But though Maxwell gave a sketch of the history of the relevant physical and philosophical ideas, he failed to give *the history of the problem situation, and of how this situation changed under the impact of the attempted solutions.* It is this lacuna that I am trying to fill here.[10]

Descartes based the whole of his physics upon an essentialist or Aristotelian definition of body or matter: a body is, in its essence or substance, extended, and matter is, in its essence or substance, extension. (Thus matter is extended substance, as opposed to mind which, as thinking or experiencing substance, is in essence *intensity*.) Since body or matter is identical with extension, all extension, all space, is body or matter: the world is full: there is no void. This is Parmenides's theory, as Descartes understood it. But while Parmenides concluded that there can be no motion in a full world, Descartes accepted a view that can be traced back to a suggestion from Plato's *Timaeus* according to which motion is as possible in a full world as it is in a bucket of water: things may move in a full world by pushing one another round in vortices; they may move like tea leaves in a tea cup.

In this Cartesian world, all causation is action by contact: it is push. In a *plenum*, an extended body can only move by *pushing* other bodies. All physical change must be explicable in terms of clockwork mechanisms (or vortices) in which the various moving parts *push* one another along. *Push is the principle of mechanical explanation.* There can be no action at a distance. (I may mention in parentheses that Newton himself regarded action at a distance at times as an absurd idea, and at other times as a supernatural phenomenon.)

This Cartesian system of speculative mechanics was criticized by Leibniz, on purely speculative grounds. Leibniz accepted the Cartesian fundamental equation, *body* = *extension*. But while Descartes believed that this equation was ultimate and irreducible, self-evident, clear, and distinct, Leibniz questioned all this: if a body pushes another body along instead of penetrating it, then this

[10]For many years I was in the habit of giving an outline of this story in my lectures. [See Popper: 'Philosophy and Physics', *op. cit.*, and *The Self and Its Brain*, 1977, especially chapters P1, P3, and P5. Ed.]

can only be because they both *resist penetration*. This resistance is *essential* to matter or body; for it enables matter or body to fill space, and thus to be extended in the Cartesian sense.

According to Leibniz, we must explain this resistance as due to forces: the thing has 'a force and an inclination, as it were, to retain its state, and . . . resist the cause of change'.[11] There are forces that resist interpenetration: repelling or repulsive forces. Thus body, or matter, in Leibniz's theory, is *space filled by repulsive forces*.

This is a programme for a theory that would explain both the Cartesian essential property of body—extension—and the Cartesian principle of causation by push.

Since body or matter or physical extension is to be explained as due to forces filling space, Leibniz's theory is a theory of the *structure* of matter, like atomism. But Leibniz rejected atoms (in which he had believed when young). For atoms, at the time, were taken to be very small bodies, very small bits of matter, very small *extensions*. Thus the problem of extension and impenetrability was precisely the same for atoms as for larger bodies. Accordingly, atoms—extended atoms—cannot help to *explain* extension, the most fundamental of all the properties of matter.

In what sense, however, can a part of space be said to be 'filled' by repulsive forces? Leibniz conceives these forces as emanating from, and—*in this sense only*—as located in, unextended points, the '*monads*': they are central forces whose centres are these unextended points. (Being an intensity attached to a point, a force may be compared to, say, the steepness of a curve at a point, that is, to a 'differential': forces cannot be said to be extended any more than differentials, though their intensities may of course be measurable and expressible by numbers; and being unextended intensities, forces cannot be 'material', in the Cartesian sense.) Thus an extended piece of space—a body in the geometrical sense—may be said to be 'filled' by these forces in the same sense in which it may be said to be 'filled' by the geometrical points or 'monads' that fall within it.

For Leibniz, as for Descartes, there can be no void: empty space would be space free of repulsive forces, and since it would not resist occupation, it would at once be occupied by matter. One might

[11]*Philosophische Schriften*, ed. Gerhardt, 2, p. 170.

describe this theory of the diplomat Leibniz as a political theory of matter: bodies, like states, have borders or limits which must be defended by repulsive forces; and a physical vacuum, like a political power vacuum, cannot exist because it would at once be occupied by the surrounding bodies (or states). Thus we might say that there is a general pressure in the world resulting from the action of the repulsive forces, and that, where there is no movement, there must be a dynamic equilibrium due to the equality of the forces present. While Descartes could explain an equilibrium only as mere absence of movement, Leibniz explains equilibrium—and also the absence of movement—as dynamically maintained by equal and opposite forces (whose intensity may be very great).

So much for the doctrine of point-atomism (or of monads) which grew out of Leibniz's criticism of the Cartesian theory of matter. His doctrine is clearly metaphysical; and it gives rise to a metaphysical research programme: that of explaining the (Cartesian) extension of bodies with the help of a theory of forces.

The programme was carried out in detail by Boscovich (who was partly anticipated by Kant[12]). The contributions of Kant and Boscovich will perhaps be better appreciated if I first say a few words about atomism in its relation to Newton's dynamics.

The no-vacuum theory of the Eleatic-Platonic school and of Descartes and Leibniz has one inherent difficulty—the problem of compressibility of bodies, and also that of elasticity. On the other hand, Democritus's theory of 'atoms *and the void*' (this was the motto of atomism) had been designed, very largely, to meet precisely this difficulty. The void between atoms, the porosity of matter, was to explain the possibility not only of movement, but of compression also. But Newton's (and Leibniz's) dynamics created a new grave difficulty for the atomistic theory of elasticity. Atoms were small bits of matter, and if compressibility and elasticity were

[12]Boscovich's *Theoria Philosophiae Naturalis* was first published in 1758 in Vienna; Kant's *Monadologia Physica*, in 1756 in Königsberg. Thirty years later Kant repudiated part of his monadology in his *Metaphysical Foundations of Natural Science*, 1786. Though the essential idea of Boscovich's monadology is to be found in Kant (see Kant, *propos.* iv and v for the *finite* number of discrete monads present in finite bodies, and *propos.* x for the central forces which are attractive over long distances and repulsive over short distances, and for Kant's explanation of extension), Kant's work is extremely sketchy as compared with Boscovich's. [*Cp.* Popper's discussion of Boscovich in *The Self and Its Brain*, 1977, pp. 190–92. Ed.]

to be explained by the movement of atoms in the void, atoms could not, in their turn, be compressible or elastic: they had to be absolutely incompressible, absolutely hard, absolutely inelastic. (This is how Newton conceived them.) On the other hand, there could be no push, no action by contact, between inelastic bodies according to any dynamic theory which—like that of Newton or Leibniz—explained forces as proportional to *accelerations*. For a push given by an absolutely inelastic body to another such body would have to be *instantaneous*, and an instantaneous acceleration would be an *infinite* acceleration, involving infinitely large forces.[13]

Thus only an *elastic* push can be explained by finite forces; and this means that we have to assume that push is always elastic. Yet if we wish to explain elastic push within a theory of inelastic atoms, *we have to give up action by contact altogether*. In its place we have to put *short-distance repulsive forces between atoms*, or, as it might be called, action at short distance, or *action in the vicinity of the agent*. Thus the atoms must repel each other with forces which rapidly increase with decreasing distance (and which would become infinite when the distance became zero).

In this way we are compelled, by the internal logic of the dynamic theory of matter, to admit (central) *repulsive* forces into mechanics. But if we admit these, then one of the two fundamental assumptions of atomism—the assumption that atoms are small extended bodies—becomes redundant; since we have to replace the atoms by Leibnizian centres of repulsive force, we might just as well replace them by Leibnizian *unextended points*, thus identifying the atoms with Leibnizian *monads*. It seems, however, that we must retain the other fundamental assumption of atomism: the void. Since the repulsive forces tend towards infinity if the distance between the atoms or monads tends to zero, it is clear that there have to be *finite distances* between monads: matter consists of a void in which there are discrete centres of force.

The steps described here are taken by Kant and by Boscovich.

[13]The argument is clearly stated by Kant in his *Metaphysical Foundations*, 'General Note to the Mechanic' (the last paragraph of the third chapter); see also his *Monadology*, propos. xiii, and his *New Doctrine of Motion and Rest*, 1758 (the section on the Principle of Continuity). Similar arguments may be found in Leibniz who writes (*Mathematische Schriften*, ed. Gerhardt 2, p. 145) 'for it seems that only elasticity makes bodies rebound'.

They may be said to give a synthesis of the ideas of Leibniz, and of those of Democritus and Newton. The theory is, like that envisaged by Leibniz, a theory of the *structure of matter*, and thus an explanatory *theory of matter*. Extended matter is explained by something which is not matter—by unextended entities such as forces and monads, the unextended points from which the forces emanate. The Cartesian *extension* of matter, more especially, is explained by this theory in a highly satisfactory way. Indeed, the theory does more: it is a *dynamic theory of extension* which explains not only equilibrium extension—the extension of a body when all the forces, attractive and repulsive, are in equilibrium—but also extension changing under external pressure, or impact, or push.[14]

There is another development, almost equally important, of the Cartesian theory of matter and of Leibniz's programme of a dynamic explanation of matter. While the Kant-Boscovich theory anticipates, in rough outline, the modern theory of extended matter as composed of elementary particles invested with repulsive and attractive forces, this second development is the direct forerunner of the Faraday-Maxwell theory of fields.

The decisive step in this development is to be found in Kant's *Metaphysical Foundations of Natural Science* in which he repudiates[15] the doctrine of his monadology according to which matter is discontinuous. He now replaces this doctrine by that of the

[14]It is important to realize that Boscovich's forces are not to be identified with Newtonian forces: they are *not* equal to acceleration multiplied by mass, but equal to acceleration multiplied by a pure number (the number of monads). This point has been clarified by L. L. Whyte (in a very interesting note in *Nature* 179, 1957, pp. 284 *ff.*). Whyte stresses the 'kinematic' aspects of Boscovich's theory (as opposed to its 'dynamic' aspects, in the sense of Newton's dynamics). It seems to me that Whyte's comments on Boscovich's critics are correct. (I may perhaps express this by saying that Boscovich gives an explanatory theory not only of extension and gravity but also of Newtonian inertial mass.) On the other hand, although Boscovich's forces are, as Whyte rightly stresses, from a formal or dimensional point of view *accelerations*, they are, from a physical, from an intuitive, and from a metaphysical point of view, *forces*—very much like those of Newton: they are dispositions, existing in their own right: they are the *causes* that *determine* accelerations. (Kant, on the other hand, thinks in purely Newtonian terms, and he attributes inertia to his monads; see his *propos.* xi.)

[15]See the second chapter, theorem 4, especially the first paragraph of Note 1, and Note 2. Kant's repudiation is founded on his 'transcendental idealism': he rejects the monadology as a doctrine of the *spatial structure* of *things in themselves*. (This is for him a 'mixture of spheres'—something like a 'category mistake'.)

dynamic continuity of matter (as an appearance). His argument may be put as follows.

The presence of (extended) matter in a certain region of space is a phenomenon consisting of the presence of repulsive forces in that region—forces able to stop penetration (that is to say, repulsive forces which are at least equal to the attractive forces plus the pressure at that place). Accordingly, it is absurd to assume that matter consists of the monads (from which repulsive forces radiate); for matter would be present at places where these monads are not present, but where the forces emanating from them are strong enough to stop other matter. Moreover, it will for the same reason be present *at any point between* any two monads belonging to (and allegedly constituting) the piece of matter in question.

Whatever the merits of this argument may be,[16] there is great merit in the proposal to try out the idea of a continuous (and elastic) something that consists of forces. For this is simply the idea of a field of forces, in the guise of the idea of continuous matter. It seems to me interesting that this second dynamic explanation of (Cartesian) extended matter and of elasticity was mathematically developed by Poisson and Cauchy, and that the mathematical form of Faraday's idea of a field of forces, due to Maxwell, might be described as a development of Cauchy's form of Kant's continuity theory.

Thus the two theories of Kant and the theory of Boscovich—which were the main attempts to carry out Leibniz's programme of a dynamic theory explaining Cartesian extended matter—became the joint ancestors of all modern theories of the structure of matter (those of Faraday and Maxwell, of Einstein, de Broglie, and Schrödinger), and of the 'dualism of matter and field', which, seen in this light, is perhaps not so deeply rooted as it may appear to those

[16]Like all alleged *proofs* in physics, Kant's proof is invalid, even in the form given here, which attempts to improve a little on Kant's own version. (Kant illicitly identifies 'moving', in the sense of a moving force, and 'movable'; *cf.* the penultimate paragraph of his Note to theorem 4. The ambiguity is bad, but it brings out his intention to identify the presence of a moving force with that of movable matter.)

The logical situation is, in brief, as follows. In this post-critical work, Kant uses his transcendental idealism to remove (by a valid argument, incidentally) his original objections to the doctrine of continuous matter, and he now thinks that he could *prove* continuity—by an invalid argument which, however, is interesting and important because it compelled him to push his dynamism to its very limits (beyond what he anticipated in his definitions).

171

who, in thinking of matter, cannot get away from a Newtonian or even from a crude Cartesian and non-dynamic model.

Another important influence deriving from the Cartesian tradition—and from the Kantian tradition via Helmholtz—was the idea of explaining atoms as vortices of the ether; an idea that led to Lord Kelvin's and to J. J. Thomson's model of the atom—whose experimental refutation by Rutherford marks the beginning of what may be described as the modern theory of the atomic nucleus.

One of the most interesting aspects of the development which I have sketched is its *purely speculative character,* together with the fact that these metaphysical speculations proved *susceptible to criticism*—that they could be critically discussed. It was a discussion inspired by the wish to understand the world, and by the hope, the conviction, that the human mind can at least make an attempt to understand it.

Positivism, from Berkeley to Mach, has always opposed such speculations; and it is most interesting to see that Mach still upheld the view that there could be no physical theory of matter. (Matter for him was nothing but a metaphysical 'substance' and as such redundant if not meaningless.) He upheld this view at a time when the metaphysical theory of the structure of matter had turned into a testable physical theory. (See also *Realism and the Aim of Science,* Vol. I of this *Postscript,* Part 1, section 17.) It is even more interesting, and somewhat ironical, that these views of Mach reached the peak of their influence at a time when the atomic theory was no longer seriously doubted by anybody, and that Mach's views are still most influential among the leaders of atomic physics, especially Bohr, Heisenberg, and Pauli.

Yet the wonderful theories of these great physicists are the result of attempts to understand the structure of the physical world, and to criticize the outcome of these attempts. Thus their own physical theories may well be contrasted with what these physicists, and other positivists, try to tell us today: that we cannot, in principle, hope ever to understand anything about the structure of matter: that the theory of matter must forever remain the private affair of the expert, the specialist—a mystery shrouded in technicalities, in mathematical techniques, and in 'semantics': that science is nothing but an instrument, void of any philosophical or theoretical interest, and only of 'technological' or 'pragmatic' or 'operational'

significance. I do not believe a word of this post-rationalist doctrine. No doubt we shall have to discard most of our theories many times; but it seems that we have at last found ways to the understanding of the physical world.

21. *Schisms, Programmes, and Metaphysical Dreams.*

The list of metaphysical programmes given in the last section was drawn up with two main purposes in mind. One was to shed some light on the significance of the present crisis in physical theory: the rejection of the Faraday-Einstein-Schrödinger programme has left us without any unifying picture, without a theory of change, without a general cosmology. Instead of a problem situation within a research programme, or relative to a research programme, our fundamental problem situation arises from a schism in physics—from a clash between two research programmes, neither of which seems to be doing its job.

Situations like this have of course arisen before. There was for example the schism between Cartesianism, which permits action only by contact (that is, at vanishing distances), and Newton's theory which permits action at a distance—a schism in which Newton himself took sides against his own theory.[1] But the present situation differs somewhat from all previous ones. Einstein's and Schrödinger's inspiring programme has been attacked by quantum theorists and, according to the judgement of most physicists, has been successfully killed. But those who attacked it have made hardly any attempt to replace it by a similarly powerful programme.

All this, I believe, is due to the prevailing philosophy of science—to the almost universal acceptance of instrumentalism, the theory of theories which Cardinal Bellarmino (one of the inquisitors in the case against Giordano Bruno) and Bishop Berkeley developed in order to oppose Galileo's and Newton's belief that science can search for truth. In the great fight over the fundamental issue of the rationalist tradition—*whether or not the human intellect, unaided by divine revelation, could uncover some of the secrets of our world*—most of the leaders of the quantum theory (except Einstein and Schrödinger) have taken sides with the Cardinal and the Bishop

[1]See my *Conjectures and Refutations*, Ch. 3, especially section iii.

against Galileo, Kepler, and Newton.[2] Although the acceptance of instrumentalism has not by itself led to the destruction of Einstein's and Schrödinger's programme, it has led to almost universal acquiescence in the absence of any alternative programme. Indeed, with Bohr's so-called 'principle of complementarity', instrumentalism proudly announces the 'renunciation' of any such programme. What is left is a tinkering with the formalism—admittedly a legitimate and even a necessary part of the method of trial and error, but only a part of it, and one unlikely to yield important results in the absence of a coherent attempt to understand the world.

David Bohm's heresy does carry on something like a part of the Einstein-Schrödinger programme; I have in mind his attempt to revive de Broglie's pilot-wave theory. But in spite of Bohm's realist and objectivist programme, his theory is unsatisfactory from the point of view presented in this *Postscript*. It is not only bound, like all other deterministic theories, to interpret probabilities subjectively, but it even retains Heisenberg's 'interference of the subject with the object'—although it tries to interpret this interference objectively. As a consequence, Bohm's reply to the Pauli-Einstein criticism (first proposed by Pauli against de Broglie's pilot waves, and later by Einstein against Bohm) seems to me utterly unsatisfactory and unacceptable.[3]

Three fundamental issues have led to the schism:

1. Indeterminism *versus* Determinism.
2. Realism *versus* Instrumentalism.
3. Objectivism *versus* Subjectivism.

The third issue arises, more especially, in connection with Heisenberg's indeterminacy relations and such questions as the reduction of wave packets; and more generally, with respect to the interpretation of probability.

It will hardly be necessary to repeat which sides are taken by the different parties over these three issues.

Einstein, de Broglie, Schrödinger and Bohm are determinists and realists; they are objectivists with respect to the aims of physical

[2] I have described this story in the chapter just quoted. [See also *Realism and the Aim of Science*, Vol. I of the *Postscript*, Part 2, sections 10–14; and section 2 of the present volume. Ed.]

[3] *Scientific Papers Presented to Max Born*, 1953; cp. *L.Sc.D.*, Appendix *xi.

theory, but subjectivists (more or less consistently so) with respect to the interpretation of probability theory.

The orthodox Copenhagen school represented by Bohr and Heisenberg and supported by Pauli, and perhaps to a lesser degree by Born, is indeterminist and instrumentalist, although all its representatives mentioned have a number of anti-instrumentalist remarks to their credit. What is most characteristic of the attitude of this school—one of its *eigenstates* as it were—is an oscillation between the objective and the subjective approach in which, owing to a kind of resonance, all its members share.

My own view is that *indeterminism is compatible with realism,* and that the realization of this fact makes it possible to adopt a consistently objectivist epistemology, an objectivist interpretation of the whole of quantum theory, and an objectivist interpretation of probability.

Although I dislike the subjectivist strain in the orthodox interpretation, I am in sympathy with its rejection of the determinism of Einstein, Schrödinger and Bohm, and with its rejection of *prima facie* deterministic theories in physics; and I agree with the substance (though hardly with the form, or with the prophetic style— the style of historical determinism) of a passage of Pauli's, taken from a letter to Born, in which Pauli rejects the deterministic research programme in the following words: 'Against all retrograde efforts (Schrödinger, Bohm, et al., and, in a certain sense, also Einstein) I am certain that the statistical character of the ψ-function, and thus of the laws of nature—which you have, right from the beginning, strongly stressed in opposition to Schrödinger—will determine the style of the laws for at least some centuries. It is possible that later . . . something entirely new may be found, but to dream of a way back, back to the classical style of Newton-Maxwell (and it is nothing but dreams which those gentlemen indulge in), that seems to me hopeless, off the way, bad taste. And we could add, "it is not even a lovely dream".'[4]

[4]Quoted from Max Born, 'The Interpretation of Quantum Mechanics', *British Journal for the Philosophy of Science* 4, 1953, p. 106. The words replaced here by dots (because they do not seem to be relevant to my discussion at this stage, any more than the historicist and determinist prophecy contained in the words 'for some centuries') are as follows: 'for example in connection with the processes of life'. (*Cp.* section 28, below.) In the *British Journal*, there is an obvious misprint ('Bohr' instead of 'Bohm').

I propose to disregard (or to shelve, 'for some centuries') Pauli's somewhat ungentle and in my opinion 'retrograde' or, more precisely, nineteenth-century historicist[5] manner of expressing his evaluations. For there seems to me much to admire in this passage. I greatly sympathize with his rejection of the Faraday-Einstein-Schrödinger programme (which he somewhat insensitively calls 'the classical style of Newton-Maxwell') on the grounds of its adherence to *prima facie* determinism, and I sympathize even more with his plea for the admission of probabilistic theories and interpretations (even though he describes them as 'statistical'). And I have no objection whatever to his description of metaphysical research programmes as 'dreams', in the sense of wish-dreams; for they are attempts to formulate our hopes, our anticipations, and our ambitions, concerning the growth of our knowledge. Yet the manner in which Pauli speaks here of these dreams betrays an ambivalent attitude towards them. There seems to be a slight but interesting change of attitude between the first two occurrences of the word 'dream' in this passage and the last. If I am not mistaken, these first two occurrences indicate something like that anti-metaphysical and anti-realistic (and thus instrumentalistic) attitude which I have described by the phrase 'tough-and-no-nonsense'. But there is a subtle change when Pauli adds, 'it is not even a lovely dream'. This, it seems, expresses two different feelings. It not only expresses the feeling that Parmenides's metaphysics—the dream of the block universe—is no longer a research programme that can attract and inspire, but also, if I am not mistaken, the longing for something better, the wish to possess a metaphysical picture of the world that is attractive and inspiring.

With such an attitude I entirely agree. Attractive and inspiring as the Einstein-de Broglie-Schrödinger programme was in its grand intuitive conception—the explanation of matter, and all its interactions, in terms of fields, and of disturbances in fields—there is something wrong and inadequate about metaphysical determinism with its block universe. And though it was a lovely dream—and one which once greatly attracted Pauli—it seems almost unavoidable

[5]See my *Poverty of Historicism*, and my *Open Society*; see also *The Open Universe: An Argument for Indeterminism*, Vol. II of the *Postscript*, sections 20–24.

that those who have breathed the freer air of indeterminism should no longer be satisfied by it: Pauli may now well say of it that it is no longer a lovely dream, or a hopeful one.

As to Pauli's dreams of the future of science, we can only guess what they are like. He has expressed his concern about contemporary science and its failure to present a comprehensive and unifying picture of the world.[6] From this remark and from his belief in indeterminism, and in probabilistic laws of nature, we may perhaps guess the outlines of a picture which might be acceptable to him.

In this metaphysical epilogue it is, I frankly admit, 'nothing but dreams' I wish to 'indulge in'. Although the dreams to be described are my own, I hope that in their general tendency they do not differ too much from Pauli's. I suppose that they would differ very little if only Pauli could forget, at least in his dreams, the instrumentalist and subjectivist element of the orthodox creed; that is, if he could forget 'complementarity' or, what amounts to the same, if he would remember that *it is likely that the world would be just as indeterministic as it is even if there were no 'observing subjects' to experiment with it, and to interfere with it.* What I take to be Pauli's main points—indeterminism and the probabilistic character of the laws of nature—will be found to be fully represented in my metaphysical dream. At the same time, and without becoming incoherent, my dream also embraces Faraday's, Einstein's and Schrödinger's programme of a physical reality that is determined by *prima facie* deterministic laws; a reality which I here take to consist of propensities. This makes it possible to join these two views, indeterminism and determinism, in a most natural manner by a correspondence argument (whereby deterministic theories are shown to be approximations of indeterministic theories); and it suggests, at the same time, a theory of matter which explains particles in terms of field concepts.

I will now try to give as rational an account of my metaphysical programme as I can, in the space which remains at my disposal.[7]

[6]*Cp.* W. Pauli, *Studien aus dem Jung-Institut* **4**, 1952, pp. 109–194, especially the end.

[7]I originally wrote this, in 1954 or 1955, intending that it should become one of the 'New Appendices' to *L.Sc.D.*, and very little room was left.

22. *Classical Determinism Modified by a Correspondence Argument.*

Let us begin by visualizing our changing world, as it changes its state from instant to instant. The states that belong to consecutive instants are, in some way or other, closely connected. This is why our world exhibits some degree of order rather than complete disorder; why it is a cosmos rather than a chaos. But we will visualize the connection between the instantaneous states not as a deterministic connection, but as something in between a deterministic block universe and chaos.

In order to get this point a little clearer, let us imagine that we have attached a film strip to a certain given instantaneous state of the world, that is, to some given 'time-slice' of the world, and that we use this film strip to represent all the past and future time-slices of the world, as well as we can represent them. (*Cp. The Open Universe: An Argument for Indeterminism,* Vol. II of the *Postscript,* section 26.)

We will assume, as a first step or first approximation, that the film strip which we have attached to the given time-slice is a Laplacean or deterministic film strip, or in other words, that it represents a deterministic block universe—the one which is determined by the given instantaneous state or time-slice; for we know that one instantaneous state or time-slice is sufficient, according to Laplace, to determine all the past and future states or time-slices of a deterministic universe. Each of the instantaneous states or time-slices of the deterministic universe we assume to be represented by one of the stills of which the film strip is composed.

Now let us further assume that we have attached a similar Laplacean or deterministic film strip to each of a considerable number of consecutive time-slices of our real, non-determinist universe.

To say that our own universe—the real universe—is non-determinist is to imply that the Laplacean film strips which we have attached to the various time-slices do *not exactly* represent our own universe; for if they did, then our real world would be Laplacean and deterministic, contrary to our assumption. No doubt the first few stills that come before and after the real time-slice to which the film is attached will be very similar to the real states or time-slices

they are supposed to represent: this we know from the success of classical physics. But minor differences will accumulate if we move further and further away from the time-slice of the real world to which the film is attached; and if we move far enough in this way, then the stills will become useless for the purposes of prediction.

Were our real world deterministic, then each film strip would represent it completely. Moreover, all the film strips would be exactly alike (since each would represent the same course of the world). But since we assume that our world is not deterministic, all, or almost all, of the imaginary film strips—the film strips which we attached in our imagination to the various time-slices of the real world—will differ from one another. Admittedly, any two of the imaginary films which are attached to neighbouring time-slices will be very similar; but they will not, as a rule, be exact replicas of each other—unless, indeed, the second of the two time-slices were *exactly* as predicted by a determinist physicist, on the basis of the first.

So far I have operated with two assumptions, and I have illustrated them with the help of our imaginary film strips (which I shall soon need for carrying the argument further): the assumption that our world is indeterministic, and the assumption that there exists a deterministic physical theory (for example 'classical physics') which is successful in the sense that its predictions are good approximations to the truth.

Now we turn for a moment from the classical extreme to the other extreme—to the assumption that the world is completely chaotic; and we try to attach another sequence of imaginary film strips to the sequence of the real time-slices, representing this assumption. How could we do it? The assumption that the world is chaotic clearly does not allow us to make any prediction. Consequently the film strips will be void of any definite information. They will leave open *all possibilities*, they will permit any state to be followed by any other (logically) possible state. Assuming that we know somehow what would be a (logically) 'possible state' of the world, each of our film strips would have to consist of stills, incorporating a catalogue of all possible states of the world, and attributing equal weight (or probability) to each possibility. Consequently all the film strips attached to the various time-slices would be exactly alike (as they

179

would be in a determinist world); moreover, all the stills of each of the films would be exactly alike, as they would be if we were living in a world without any change in time: for there is only *one* complete catalogue of possibilities. (This catalogue can be presented in various ways, for example by 'possibility spaces'—abstract multi-dimensional spaces such that certain three-dimensional sub-spaces each represent a possible arrangement, or a configuration of particles; or that each point—or vector—represents one of the possible states of the physical system in question which, according to our assumption, is the system of the world.)

Returning to the deterministic or 'classical' film strips we may now, if we like, estimate of each still how similar it is to the time-slice it is supposed to represent. For example, we may measure the deviation of every represented detail from the real state of the world which it is supposed to represent. To this end, we should have first to agree upon some definition of an 'elementary detail' or 'event'. (It must not, roughly speaking, be too small a detail, since the assumption that classical physics is applicable down to *every* structural detail of the world may lead to trouble; consequently our classical film cannot contain a very detailed description. However, it is not necessary for our purpose to say more about this point here.) In addition, we may agree on some yes-or-no criterion by which we decide whether or not an elementary event is properly represented in our still. We can then estimate the average number of the 'yes' answers, or in other words, the chance, or the probability, of obtaining from the still in question a correct elementary prediction. This probability may then be taken as a measure of the similarity between the still and the real state of the world. The probability will be almost 1 for stills close to the time-slice to which the film strip is attached, and it will show a general tendency to decrease if we move further and further away from this instant into the 'past' or the 'future' of the film. The point at which the film becomes completely useless for predictive purposes will be reached as soon as we have to assign to a still the probability ½ or less; for this means that a random answer to any yes-no question about the world will be as good as an answer based upon our classical film. From this still on, we know that, even if the degree of correspondence between a still and a represented time-slice should be greater than ½, this can only

be due to accident; thus from this still on, the classical film will be just as useless for predictive purposes as the catalogue of possibilities which represents the other extreme, the hypothesis of a chaotic world.

Note that we could not, of course, ever *determine in advance* the probability here described of future stills: since we do not know in advance what the real world will be like, we cannot determine the number of events in which a predictive still agrees with the time-slice which it represents. We should therefore have to be content to ascribe something like generally declining probabilities to the stills, from 1 down to ½, according to their temporal distance from the time-slice to which their film is attached. The rate of this decline could be determined only by averaging over past correspondences. (There is no need, however, to go more deeply into this matter, since we shall replace those probabilities which we have assigned to our deterministic film strips by a completely different probability which will be built in, as it were, into a kind of indeterminist and non-classical film strip, to be described in the section after the next.)

I have tried to elucidate some aspects of the relationship between our indeterminist real world and its determinist or classical representation. It turned out to be necessary to operate with a sequence of film strips, or with a *multiplicity* of classical representations, ever to be renewed. This first modification of classical physics was an immediate result of an argument according to which classical ideas cannot be more than approximations to a better theory, and to the truth, since no scientific representation can completely predict an indeterministic world.

In agreement with most other interpretations, I here conceive classical physics as an approximation to quantum physics. (This is the 'correspondence argument'; *cf. Realism and the Aim of Science*, Volume I of the *Postscript*, section 15.) Yet there is a fundamental difference between this way of seeing these things and the orthodox interpretation. We take indeterminism as a cosmological fact which we do not attempt to explain. But the orthodox interpretation tries to explain this fact as due to our own interference with the physical process: as if the world would be deterministic (or more deterministic than it is now), if only there were no interfering men about; as if

181

quanta (like some children) would behave in a more orderly or predictable fashion, if only nobody was looking. This view seems to me absurd; in order to make it more acceptable, the orthodox interpretation is forced into an idealistic or semi-idealistic attitude towards the world—into an attitude which makes it meaningless or semi-meaningless to speak of a reality which is there when nobody is looking. But no such *ad hoc* philosophical assumption is needed. The situation is as simple as it can be. I will try to show that this is so.

Bohr has always contrasted the natural, the intuitive, the immediately understandable character of classical physics with the more difficult, non-intuitive, and highly sophisticated character of quantum physics; and he has suggested that we may be able to make quantum physics more understandable to ourselves if we make use of correspondence arguments, that is to say, if we keep before our minds the transition from the quantum theoretical case to the classical case (which latter we can understand intuitively). Now I have made use here of a correspondence argument for the purpose of sketching the outlines of the indeterminist's picture of the world; yet I wish to draw the reader's attention to the fact that the resulting non-classical picture is more natural, more in keeping with ordinary experience, and less sophisticated, than the classical view of Hobbes, say, or Laplace.

For assume that we know today's time-slice, or part of it, and that we wish to predict certain events a year ahead. It is, I think, just what we should 'naturally' expect to hear if we are told that certain events, such as the flight of a bee, or the movement of a cloud, are not yet predictable—although they may become so, more or less, if we have perfect knowledge of a time-slice *very shortly ahead of them*; if we are told, further, that other events, such as solar or lunar eclipses, can be predicted far in advance; if we are told, in general, that the knowledge of a time-slice six months ahead of an event would help us to foresee some of its details a little better than knowledge of a time-slice a year ahead of that event. And this is precisely my point. What I have attempted to convey with the help of my picture is just this: that indeterminism forces us to adopt the view that there can be no theory which completely predetermines all events ahead; that therefore each time-slice yields its own predictive film which, however, soon loses its usefulness; that later and later time-slices give us a better and better idea of an event ahead of them

all; and that we therefore have to try, if we want a detailed and reliable prediction, to obtain a description of a time-slice (or part of it) of a recent date—as recent as possible.

All this is very simple; and it strikes me, intuitively, as more natural and familiar than the classical Laplacean dream of unlimited predictability. It is, in a way, too simple; and I shall soon (in the section after the next) enrich my picture by replacing my deterministic or classical film strips by others: by film strips that describe propensities. Yet even in the simple version given in the present section, our picture contains the whole story of the 'reduction of the wave packet': the transition from one basis of information to another—a later, and thus a fuller and a better one. (And if we were to believe with Heisenberg that this 'reduction' is the same as 'the element of discontinuity in quantum theory', then we should have to say that our very simple and primitive picture already incorporates quantum discontinuity, that is to say, the idea of a 'quantum jump',[1] which only shows how little this latter-day quantum jump has in common with the original idea.)

Before proceeding to replace our classical film strips by non-classical ones, I will show next how the 'reduction of the wave packet' appears here as a consequence of the hypothesis of an indeterministic world for which classical physics holds only approximately.

23. *Indeterminism and the So-Called 'Reduction of the Wave Packet'*.

The metaphysical view which I am trying to sketch seems to me in many respects similar to the orthodox view. It is for this reason important to appreciate clearly those of its tenets which differentiate it from the orthodox view. I mean its realism and its objectivism. This will be done by a further discussion of the picture which has been introduced in the foregoing section, and which consists of a sequence of time-slices of the real universe, and a sequence of imaginary classical film strips, each of which is attached to, and determined by, one of the time-slices.

The succession of actual time-slices is an objective, real process, possibly a continuous one. The film strips are of course imaginary;

[1] *Cf.* note 6 to section 13, and text.

but they are objective in the sense that each of them is logically entailed, as it were, by a complete description of the time-slice to which it belongs (in conjunction with a complete deterministic or 'classical' system of physics).

The question may be raised whether the change from one imaginary film strip to the next can be represented by a continuous function of the changing time-slices, provided the latter do change continuously. Clearly, our assumptions so far developed here do not suffice to decide this question; but the quantum theory seems to imply that one imaginary film strip will have to be replaced by another whenever some interaction has taken place, such as a collision, or an emission, or an absorption, or any other form of energy exchange. Thus in a world consisting of free particles at great distances apart, *copies of the same classical film strip might perhaps be attached even to time-slices which are removed in time from one another*. (This seems feasible in view of the validity of the classical conservation law of momentum.) But in a dense and complex world system, *new* films would have to be attached after extremely short time intervals, since new and unpredictable states would be realized with every interaction.

Now we have seen that a prediction—a prediction, say, of an event on the first of January of the year 2000—will, in general, be the more definite and the more reliable the nearer to the event in question we can choose the (total or partial) time-slice upon which our prediction is based. So if we are really interested in knowing something about this event, we shall try, from time to time, to bring our information 'up to date'; that is to say, to obtain information about a more recent time-slice, in order to base our prediction upon the latest available initial conditions. We may do this even in a deterministic world, in view of the limited precision of our measurements; but in an indeterministic world we should have to do so, even on the assumption that our information concerning the initial conditions was absolutely exact and complete.

Now every new prediction based upon new and later information of this kind will differ somewhat from the previous prediction. This does not mean that the previous prediction will have been wrongly calculated. It only means (i) that we have to consider the prediction as relative to the time-slice, or information, on which it was based,

and (ii) that it should be clear to us that the use of a later time-slice
will in general improve the value of the prediction based upon it.

The logical situation here is precisely the same as that which we
encountered in our analysis of the so-called 'reduction of the wave
packet'. For let e be the event whose presence or absence we wish to
predict, and let s_1, s_2, \ldots be classical film strips attached to later and
later time-slices. (All time-slices considered will be assumed to be
prior to the event e.) Let

$$pred\ (e,\ s_1)$$

be a prediction with respect to e in the light of the appropriate still of
the strip s_1. We shall then find that $pred\ (e,\ s_1)$ and $pred\ (e,\ s_2)$ do not
in general agree, and that the latter will generally be preferable as a
prediction to the former.

The transition from $pred\ (e,\ s_1)$ to $pred\ (e,\ s_2)$ corresponds exactly
to the transition from the probability statement $p(e,\ s_1)$ to $p(e,\ s_2)$
where '$p(a,\ b)$' denotes the probability of a given the information b.
But a transition from $p(e, s_1)$ to $p(e,\ s_2)$ is, as we have seen, precisely
what quantum theorists have called a 'reduction of the wave packet'.
(See section 8 above.) They have suggested that this reduction of the
wave packet is connected with, or dependent on (a) the measuring
experiment by which we obtain new information s_2, and (b) the
realization or actualization of what so far was only potential.
(Heisenberg's 'transition from the possible to the actual'; see note 3
to section 9, and sections 10 and 13 above.) These two points, (a)
and (b), are often combined in the suggestion (c) that it is only under
the stimulus of our own interference with the physical system, only
owing to our measuring experiment, that the transition from the
possible to the actual takes place. In our picture, by contrast, the
transition from the possible to the actual takes place whenever a new
state of the world emerges; whenever a new time-slice is actualized
or realized, whether observed, or measured, or not. (In fact,
observations and measurements are so extremely rare that almost all
'actualizations of potentialities' happen independently of them.) As
long as anything happens, as long as there is any change, it will
always consist in the actualization of certain potentialities. Thus a
new film strip (and with it, an opportunity for the reduction of the

wave packet) appears whenever any interaction takes place. Whether or not we *know* or *observe* the new state s_2, and whether we replace *pred (e, s_1)* by *pred (e, s_2)*, in our attempts to predict *e*, is completely incidental, and does not in any way bring about the actualization of potentialities. *The world changes without reference to us.* We can be quite sure that many interactions have occurred between our choice of strip s_1 and our choice of strip s_2, and that these interactions were not influenced by our choices.

Of course, some changes *are* due to our own experiments; and these are both practically and theoretically important to us. But it looks to me very much like a symptom of either myopia or megalomania to allow one's view of the world, or of science, to be dominated, or even coloured, by the disturbances created by one's own experiments. Transitions from the possible to the actual and quantum interactions were going on before anybody ever interfered with anything, and they will continue to go on long after we have all left off interfering.

24. *Replacing the Classical Films by Propensity Films.*

So far our modification of the classical picture has been somewhat crude: we replaced the one imaginary classical film strip by a sequence of many different film strips, each of them attached to a time-slice of the real world. If we wish to refine this picture, we may consider that so far we have made no use of the other extreme, as I called it—the representation of chaos by a catalogue of all possible states. Let us assume, then, that we possess this catalogue, and a film strip such that every still of it consists of this catalogue. Our problem is to find a sequence of film strips of a new type, somewhere in between these two extreme sequences of film strips, the classical and the chaotic type. The new type should of course give us more information than either the classical or the chaotic type.

The solution of this problem (suggested by quantum theory) is this: the new type of film strip will consist, like the chaotic type, of catalogues of possibilities; but to each of these possibilities will be ascribed a probabilistic measure or *'weight'*. This turns the catalogue of possibilities into a probabilistic distribution—a distribution of properties.

It will be remembered that in one of the two extreme cases discussed in section 22—in the case of complete chaos—we found that all the film strips became identical, each being simply a repetition of the same complete catalogue of all possibilities. But once we ascribe measures or 'weights' to the possibilities, there will be an immense number of different complete catalogues, each with a different distribution of weights over the various possibilities. It is further clear that if we consider one of these weighted catalogues that replaces, say, a classical still not far removed in time from the real time-slice to which the film is attached, then it will have to give most of the weight to those of the possible states which are very similar to the state which the classical theory predicts; consequently, it will give hardly any weight to most of the other states. But this implies that, in any of the films, the distribution of weights in consecutive 'stills' (that is, catalogues) will be closely connected; or that the distribution of weights in one moment will *determine* the distribution of weights in the next moment. As a consequence, *the succession of stills (of weighted catalogues) in each film will be of a deterministic character*—very much like the succession of stills in the deterministic or classical film strips which we considered first (in section 22). The difference will be this: laws of a determinist character (although not necessarily identical with the classical laws) now connect *catalogues of the 'weights' of all possible states of the world,* while before, the classical laws connected a representation of one state of the world with that of another state of the world.[1]

The character of the new laws will be deterministic (and thus 'quasi-classical') simply because the probability distribution of the later still, or catalogue, must depend on, or be a function of, the immediately preceding still (which is also a catalogue). And this means no more than that some of the states are never, or hardly ever, followed by some other states: even though these other states are logically possible, they are excluded by the laws—they are forbidden to succeed the preceding state. Thus the laws that determine the

[1][*(Added 1981) Since each 'still' in the filmstrip attached to a real time slice consists of a whole catalogue of weighted possible states, my proposal really involves that the predictive filmstrips split at any interaction into as many filmstrips as there are possibilities in the catalogue. In this my picture—this section was written in 1954 or 1955—greatly resembles Everett's; only that the many worlds remain mere possibilities instead of becoming real. See my discussion of Everett in the 'Introduction' above, section 5.]

weights, or the distribution of the propensities, by connecting each still (or catalogue) with the next one, will be laws of a deterministic or near-classical character (like the time-dependent Schrödinger equation).

I believe that this new picture, though perhaps a little complex, adequately represents the status of quantum theory, and its relation to classical theory.

Yet what is the advantage of this more complex picture over our first modification? What is its advantage, more especially, over our earlier procedure of attributing probabilities to the classical stills, in order to remind us of their decreasing usefulness? One advantage can be seen at once. Assume that the classical still that represents the world one week ahead has obtained, on the basis of our first procedure, a probability of 0.7 (which means here an *average rate of success* of similar predictions made one week ahead in the past), and consider *(a)* a weather forecast based upon it, and *(b)* an eclipse prediction. Clearly, the weather forecast will be less reliable and the eclipse prediction more reliable than this average rate of success indicates. Our new method can fully differentiate between these cases: a possible event—as opposed to a possible state of the world—will obtain a probability equal to the sum of the probabilities of all the possible states which contain this event. Thus if we succeed with our idea of attributing weights to all possible states, then the probability of the eclipse, or of its non-occurrence, can become much higher, and that of a certain weather situation, or of its non-occurrence, can become much lower, than the probability of 0.7, attributed to the classical still in question. This shows that the new modification may yield better results than the old one.

Now let us consider some of the problems raised by our new picture—the replacement of the classical film strip by a catalogue of weighted possibilities. The main problem is, of course, the determination of these weights or probabilities; or in other words, the discovery of the laws which allow us to connect one still—that is, one weighted catalogue, one probability distribution in possibility space—with the next. This problem sets us the task of determining transition probabilities by a generalization of the classical dynamical laws of nature.

This is the task which quantum theory attempts to carry out.

Quantum theory, as Einstein once said, 'dethroned classical physics as far as application to the case of sufficiently small masses is concerned. . .; so that today the laws of motion formulated by Galileo and Newton can be considered valid only as limiting cases'.[2] So much was clear even before 1925. The problem was, what could be preserved of the old dynamical laws? The theory of Bohr, Kramer, and Slater, of 1924, based on intuitive ideas to which my present picture is in many ways indebted, assumed that even the conservation laws of momentum and energy were only valid for statistical averages. This view was refuted by experiments of Bothe and Geiger; and so the new quantum theory of 1925–26, linked with the names of Heisenberg, Born, Jordan, Dirac, and Schrödinger, was developed, which satisfied the classical conservation laws.

Moreover, Schrödinger's theory was 'classical' in the sense that, *with respect to the propensities,* it was *prima facie* deterministic: it *determined* the distributions of the propensities; and it was time-reversible. From the present point of view this is almost to be expected. For we are faced with the following problem: (i) We have to find laws which play a part analogous to the classical dynamical laws of nature, but which connect distributions of propensities rather than actual classical states. This suggests that the laws should be, if possible, differential equations, determining changes of densities, or of weights, of continuous (multi-dimensional) fields of propensities. (ii) We have to find laws which have a high degree of simplicity or testability; but we know already that laws of the classical *prima facie* deterministic type are simpler, and thus better testable, than other laws. (iii) Our demand for *indeterminism* is satisfied by attaching different film strips to different time-slices, and by replacing the classical films by probabilistic propensity films; thus there is *no need* for the dynamic laws that are to connect the stills to be further removed from the classical type. (iv) This is also suggested by the demand that classical physics should be an approximation of quantum physics (that is, by the principle of correspondence; *cf. Realism and the Aim of Science,* Volume I of the *Postscript,* Part 1, section 15).

But let us return to the conservation laws. Their validity for individual processes suggests that non-interacting particles (and *a*

[2]A. Einstein, *Mein Weltbild,* 1934, p. 173. (*Essays in Science,* p. 8.)

fortiori if these are unobserved particles—but one of my main objects here is to brush aside all problems of observationalism) behave classically: a view held by Einstein. It is the interaction of particles—including photons—that is indeterministic, and especially the interaction between particles and particle structures such as screens, slits, grids, or crystals. These interactions preserve the conservation laws; *but the conservation laws do not suffice for determinism.* This is so because particles may be conceived, not as billiard balls, but as carriers with probabilistic propensities to interact (with various biases). Thus we assume, with Dirac, that a particle approaching a polarizer has a certain propensity to pass it, and a complementary propensity not to pass it. (It is the whole arrangement which determines these propensities, of course.) There is no need to attribute this indeterminism to a lack of definiteness or sharpness of the state of the particle, or to the indeterminacy relations: these arise, rather, as scatter relations, in consequence of the fact that deterministic interaction is replaced by propensities. This view not only replaces the mistaken belief that the indeterminacy is 'due' (or rather partly 'due') to our interference, to our measurements, etc., but it also explains it, up to a point. For every measurement is based upon the interaction of particles; and it will therefore indeed create scatter, in accordance with the distribution of the propensities. But the same happens also in innumerable cases in which there is no observer and no observation.

All this shows that the situation is closely related to the soldier's random walk (discussed above in section 10). In fact we are now in a position to construct something like a highly simplified model for quantum theoretical indeterminism, by the simple device of introducing more than one soldier.

25. *A Rough Model of Quantum Theoretical Indeterminism.*

We take a very large paddock or field, surrounded by a wall, and distribute on it a group of soldiers, consecutively numbered, each equipped with a pocket roulette-wheel with a symmetrical spinning needle whose two ends are not distinguished. Each soldier is instructed to march in a straight line with constant velocity until he comes within five steps' distance of either the wall or another

soldier. If it is the wall, he continues his movement as if he were a billiard ball reflected by it. If it is another soldier, then the one with the higher number consults his roulette-wheel; and after this consultation—without loss of time—he moves off in one of the two opposite directions defined by his roulette-wheel. His direction and speed, and also the direction and speed of the other soldier, are to be so determined that together they satisfy the laws of conservation of momentum and energy; which is always possible.

Our model shows, first, that the conservation laws do not entail determinism, but leave one of the variables open to choice, or to determination by chance. (This is not so, of course, in the case of elastic impact of macroscopic billiard balls.) Secondly, it shows that, beginning with the first encounter of two soldiers when the original predetermination of their movements is destroyed, we have to operate with catalogues of possibilities which may be graphically represented as spreading waves (*cf.* section 10 above).

A third point is this. Let us assume that in order to determine a soldier's position, it is necessary to arrange an encounter with him. Then our model shows that every determination of the position of a soldier will interfere in an unpredictable way with his momentum. Admittedly, this third point is still *far removed from Heisenberg's indeterminacy relations;* but it indicates the way in which the interference of the observer with the object may be visualized as *something of no fundamental importance*—a *consequence* of the prevailing laws of interaction, and therefore in no way capable of explaining these laws, or their indeterminist character.

A fourth point illustrated by our model is this. Let us consider one of two soldiers who are about to meet, and let us try to calculate his propensity to be located, ten minutes after that encounter, at a certain given region of the field. It is clear that this propensity will depend on the *total situation*: all soldiers so positioned that they may possibly encounter our soldier within these ten minutes will influence the result of the calculation. The changing possibilities, or propensities, of their locations will all influence the propensity which we are trying to compute, and will interact with it.

Our example is, of course, much too simple to produce a .case analogous to the two-slit experiment. Yet it can be used to illustrate what is meant by saying that every state of the system is a realization

of the potentialities determined by its immediately preceding state; and it may also be used to illustrate the idea that to every instantaneous state or time-slice of the system is attached an imaginary film strip whose stills are catalogues (or spaces) of weighted possibilities, and that each of these strips will be supplanted by a different one upon any interaction between the particles (or upon any realization of some of the possibilities). Of course even a supplanted film strip may still be useful for many probabilistic predictions. But if we can obtain later information, then we shall do so, and thus 'reduce our wave packet'.[1]

26. Matter and Field.

Einstein's most cherished dream for perhaps forty years was to construct a unified field theory—a field theory in which the dualism of matter and field was superseded, and in which particles were *explained* as resulting from the properties of the field. The object of a theory of this kind is, of course, the deduction of the *physical properties* of the particles—their stability or instability, their laws of motion, their mutual interaction, and their interaction with the rest of the field—from the assumed equations of the field. Einstein's

[1] It seems to me worth noting in this context that *Descartes's theory of free will* was based on his belief that although the 'quantity of motion' was preserved (no doubt an intuitive anticipation of the energy principle), its direction was not: this is why the discovery of the law of conservation of momentum was generally assumed to refute his view that the non-material mind may have power over the direction of motion (although not over its quantity). But if we bring in (at least) *two* particles, then the laws of conservation of momentum and of energy certainly do not fully determine their direction, although they do determine, after the direction of one of them has been chosen, the speed of both particles, and the direction of the second particle. Now what a moving animal does is precisely this: it chooses its direction. It can do this, in most cases. Its choice of speed, however, is usually limited: that is, by the available muscular energy, by the maximum acceleration it can develop indirectly, and thus, by the energy principle. (If we take the animal as one of these two particles, then the second particle whose presence makes it possible to satisfy the law of conservation of momentum is, as a rule, the earth.) I do not think that any physical law is violated here; nor that physical laws suffice to determine all animal movements—such as those of my hand guiding my pen. In other words it seems to me that Descartes's idea was fundamentally correct, although his physics, of course, needs amending. [*(Added in proofs.) I now find that Schrödinger said long ago (in 1922): ' . . . the energy-momentum theorem provides us with only four equations, thus leaving the elementary processes to a great extent undetermined . . .'; cf. *Science Theory and Man*, 1957, p. 143]

original programme does not seem to be regarded as very hopeful any longer by many physicists. This, I believe, is partly a consequence of Einstein's long adherence to determinism, and partly due to the fact that his idea of a field—which stems from Faraday—is different from the idea to which the quantum theory has given rise—the idea of fields that determine probabilities (to be interpreted, in my view, as propensities).

A Faradayan or Einsteinian field may be conceived as being just as 'real'—or almost as 'real'—as an electron, or a Johnsonian stone: such a field 'can be kicked, and it kicks back' (to quote Landé). Indeed, the present physical explanation of kicking and of being kicked back is (in first approximation), quite generally, that of two Coulomb fields kicking each other. Fields of probabilities or propensities are more abstract things. They can be created (by an experimental set-up) but they cannot be kicked, and they do not kick back.[1] However, an experimental kick may correspond to a set-up whose field determines a very high probability for a kick-back. Since it can do this, and since we have been led to assume that possibilities—virtual events—can influence one another and thus the final results, I propose to consider fields of possibilities or probabilities as 'real'—as manifolds of 'real', objective, though virtual, propensities. But since they cannot be kicked, and do not kick back, they will have to be conceived, somehow or other, as less 'real' and more 'abstract' than particles of Faradayan or Einsteinian fields.

Another aspect of more recent developments seems to point in a direction very different from that towards which Einstein aimed with his unified field theory. Einstein tried to construct a theory in which the two kinds of field considered at that time in physics—Faraday's and Maxwell's electromagnetic field and Einstein's own gravitational field—were replaced by one single field. This particular programme of unification was closely connected with what may be called the electrical theory of matter—the belief, in brief, that matter was a form of electrical energy.

In its simplest form, this theory tried to explain the mass of electrons and protons as due to the energy content of the electrosta-

[1][This view stems probably from 1954. It was later revised by Popper. See his Introduction to this volume, thesis 8. Ed.]

tic field surrounding them; and in this form, at any rate, the theory had to be abandoned as refuted upon the first discovery of a neutral material particle, the neutron. (It has been refuted again every time a new neutral elementary particle has been discovered. Of course these discoveries constitute refutations only if we assume that these neutral particles are 'elementary', that is, non-composite.)

With the abandonment of the electrical theory of matter, the particular programme for a unified field theory which had been envisaged by Einstein lost its plausibility. Indeed, with any new kind of particle (whether neutral or charged) the present quantum theory associates a new kind of quantized field: instead of a merger of two fields into one, we now have as many different kinds of fields as there are different kinds of particles. They numbered at least sixteen in 1957 (if we always associate one kind of field with two particles—that is a particle and its anti-particle).

In any case, the fundamental idea of a unified field theory seems to me one that cannot be given up—unless, indeed, some alternative unified theory should be proposed and should lead to success. For the present state—that of a multitude of field theories—seems unsatisfactory in several respects. Admittedly, it has led to most satisfactory quantitative predictions in electrodynamics—the probabilistic theory of the interaction of an electron-positron field (an 'electron-positron assembly') with a photon field (a 'photon assembly'). But outside electrodynamics the predictions derived from the present theory are mainly qualitative; which means that the theory cannot be satisfactorily tested.

Moreover, the situation is unsatisfactory even within electrodynamics, in spite of its predictive successes. For the theory, as it stands, is not a deductive system. It is, rather, something between a deductive system and a collection of calculating procedures of somewhat *ad hoc* character. I have in mind, especially, the so-called *'method of renormalization'*: at present, it involves the replacement of an expression of the form '$\lim \log x - \lim \log y$' by the expression '$\lim (\log x - \log y)$'; a replacement for which no better justification is offered than that the former expression turns out to be equal to $\infty - \infty$ and therefore to be indeterminate, while the latter expression leads to excellent results (especially in the calculation of the so-called Lamb-Retherford shift). It should be possible, I think, either to find

194

a theoretical justification for the replacement or else to replace the physical idea of renormalization by another physical idea—one that allows us to avoid these indeterminate expressions.

But more important than shortcomings of this sort seem to me the present methods of operating with thirty or thirty-two different kinds of fundamental particles (counting the anti-particles) with which are associated fifteen or sixteen different kinds of fields.

About the unsatisfactory character of this situation there is, I think, unanimity; also about the desirability of explaining, if possible, the thirty odd particles as states of a quantized field, and their decay properties by transition probabilities (which has been partly done). It is an aim like this that makes a unitary field theory urgent. The aim is determined by a research programme which is often implicitly accepted, although it is hardly articulated.

Of course, the usual way of looking at fields and particles is essentially dualistic; and it is widely held that this dualism is an essential feature of atomic theory. This is not surprising when we consider that the idea of 'associating' a field with a particle has proved most fruitful, from Einstein's photon theory to de Broglie's and Schrödinger's theory of the electron and to Yukawa's theory of the meson. And yet, this dualism does not seem well founded as yet, however helpful it may have proved in handling physical problems. Born's statistical interpretation, for example, is an essentially monistic particle theory (*cf.* section 11 above, text to footnote 6). As Landé points out: what exists are particles; the waves merely determine the frequency with which particles assume a certain state, upon repetition of the experiment. Similarly, the more recent theories—involving a multiplicity of quantized fields associated with the various kinds of particles (mentioned earlier in this section)—are also, it must be stressed, essentially particle theories. For, these field theories are statistical theories of *particle assemblies*: what they describe are the numbers—or more precisely, the probabilities of a change in the numbers—of the particles in the various possible states; they describe the probability of 'creation' or 'destruction' of particles in these states ('creation and destruction operators'), or, with a different choice of metaphor, the probability of the transition of particles of one state to another.

Thus whether we look at the quantum theory of Heisenberg,

Schrödinger, and Dirac, in Born's interpretation, or at the more recent theories of quantized fields, there really seems no basis for the assertion that quantum theory incorporates a dualism of particles and waves: in all these theories, the 'waves' play the part merely of determining the probabilities for particles to take up certain states, or to undergo transitions from certain states to certain other states.

Where however a genuine field theory is likely to come in (and a unified field theory), as opposed to a particle theory, or a theory of particle assemblies, is in the explanation of the particles themselves. This was first foreshadowed by Schrödinger in his original attempt to explain particles as wave crests (or wave packets); and though he has given up this theory he has often returned to the problem.

A little consideration of what a monistic field theory—or any other theory—may be able to do could help us here. As indicated earlier, it could explain, at best, only *the physical properties* of particles; or in other words, their *physical behaviour*; or still more precisely, their *disposition or propensity* to behave, under certain circumstances, in certain ways. No physical theory can do more than this: it can describe a physical system only by describing its propensities. Thus particles are propensities, from the point of view of physical theory; and it is only *a certain metaphysical view* which sees them differently—perhaps as regions of space which are, temporarily, packed 'full' of something, and indivisible.

I suggest that we give up this metaphysical view, and that we replace it by another, equally metaphysical one: by the view that propensities are real; that they are described by field equations; that particles can be produced by propensities; and that, at least up to a point, particles *are* propensities; so that they are, in this respect at least, what the physical theory tells us—what it can possibly tell us. We can in this way supersede the dualism of matter and field without sacrificing any advantages it may have for the treatment and solution of physical problems. For although we replace this dualism by a monism of propensities, we retain, within this monistic view, a kind of practical dualism. For propensities are, on the one hand, *potentialities*; and on the other hand, they are propensities or potentialities to *realize* something. But whatever may be realized, or may realize itself, must again be a set of propensities or potentialities to realize something else.

196

A physical theory of matter which treats particles in the way indicated here is not only a programme, or a wish dream, for it was achieved many years ago, as far as positive electrons are concerned. I have in mind, of course, Dirac's famous theory—one of the boldest and most ingenious parts of quantum theory. It is a theory which interprets the positive electrons as 'holes'—that is to say as unoccupied states, as open possibilities for occupation. The vacuum (the empty space) is conceived by this theory as possessing a structure: it consists of negative electrons which 'occupy' all or almost all the possible states of negative mass and energy. (The existence of such states is a consequence of Dirac's equations.) Having assumed a state of this kind, a negative electron disappears, as it were; but it remains virtually present, in so far as it may re-appear under certain circumstances if a light quantum supplies the necessary energy to lift the electron to a higher energy level. But simultaneously, a 'hole' will appear—its previous state which is now unoccupied; and the theory predicts that this 'hole' will behave exactly like an electron with a positive electric charge: it will be repelled by positive charges, attracted by negative charges, and will 'unite' with a negative electron, whereupon both the 'hole' and the negative electron will disappear. Their charges as well as their masses will be annihilated and converted into radiant energy (into a photon). The two inverse processes—the creation of a pair of positive and negative electrons out of the vacuum, and the annihilation of a positive and a negative electron—are thus explained. The explanation is monistic. There is only *one* kind of particle involved, the one which we know as the negative electron; and the two processes consist in its assumption of different states, of different energy levels. The other particle—the one we know as the positively charged electron—turns out to be a 'hole' in the vacuum: an open possibility, an unoccupied state which may be taken up by a negative electron: a *propensity*.

The particular kind of monism inherent in this approach—a monism of negative electrons—is hardly any longer attractive today. Too many different particles (*e.g.*, mesons) have been discovered, since the discovery of the positive electron, which cannot be explained in the way described. The time when there was a hope that it would be possible to explain matter in terms of the states of one or two fundamental particles has passed, it seems.

What is important here is that the existence of Dirac's formalism proves the existence of a mathematical theory which allows us to describe the presence of a material particle as equivalent to a 'hole', that is to say, to a non-particle, to a potential but 'unoccupied' state of a particle: to a sheer open possibility for interaction with other particles.

There does not seem to be any reason why these other particles should not in their turn be mathematically equivalent to open possibilities. If so, we should arrive at another kind of monism—one which interprets material particles as both the *realizations or actualizations* of certain potentialities (pair-creation realizes certain potentialities of the vacuum) *and also as potentialities for certain kinds of interaction* (such as pair annihilation, for example). A formalism which would treat 'particles' and 'holes' throughout as equivalent would, up to a point, fulfil Einstein's and Schrödinger's programme: it would explain particles of matter in terms of fields. At the same time, it would fulfil a programme which has been sketched here: a programme for a theory of change which would be monistic and dualistic at the same time; which would allow us to interpret any real state of the world as both an actualization or realization of some of the potentialities or propensities of its preceding states, and also as a field of dispositions or propensities to realize the next state. In this way, the apparent dualism of matter and field, and of particle and wave, would be shown to spring in the most natural way from the two fundamental aspects of every physical thing. I mean its two aspects as a *bearer of dispositions*—a *bearer* that can have no further testable properties beyond these *dispositions*.

A theory like this might also yield the principle that particles representing the same possibilities are identical (and with it Bose's principle).

27. *Open Problems.*

My dream programme is metaphysical. It is non-testable: it is irrefutable (and irrefutability, we should remember, is not a virtue but a vice). It is based upon the metaphysical (rather than the

'scientific') idea of indeterminism.[1] It tries to supplant, to super-sede, the existing metaphysical interpretation of quantum theory in terms of 'particles', defended by Born against Schrödinger; it also tries to supplant the instrumentalist interpretation of Bohr (who renounces any attempt to go beyond the particle-wave dualism and explain it). And it tries to give a coherent view of the physical world—a physical world which is no longer a strait-jacket for its physical inhabitants, not a cage in which we are caught, but a habitat which we may make more habitable, for ourselves and for others (and which incidentally, we are about to make uninhabitable for our children by what we proudly call 'the peaceful use of atomic energy'[2]).

But if my dream is metaphysical, what is the use of it? Is there anything in it beyond, perhaps, an emotional satisfaction? Is it not utterly different from a scientific hypothesis—one in which we are mainly interested *because of its implicit claim to be considered, tentatively, as true?*

I no longer think, as I once did, that there is a difference between science and metaphysics regarding this most important point. I look upon a metaphysical theory as similar to a scientific one. It is vaguer, no doubt, and inferior in many other respects; and its irrefutability, or lack of testability, is its greatest vice. But, *as long as a metaphysical theory can be rationally criticized,* I should be inclined to take seriously its implicit claim to be considered, tentatively, as true. And I should be inclined to evaluate it, in the main, by an appraisal of this claim—considering its theoretical interest first, and taking only a secondary interest in its practical usefulness (as distinct from

[1][See *The Open Universe: An Argument for Indeterminism,* Vol. II of the *Postscript,* for the distinction between scientific and metaphysical indeterminism. Ed.]

[2]Although the topic is a digression, we might reflect here on the fact that, some forty thousand years or more after Prometheus, we have not learned to master the dangers of 'the peaceful use of fire'. People are still killed by it, peacefully and regularly, although there is no inherent difficulty in controlling it, and although a burnt child dreads it (unlike mislaid radio-active phials and other 'atomic wastage'). The building of atom bombs may be, in an unfortunate political situation, an unfortunate necessity; but can the same be said of most of the forms of 'peaceful use' of atomic energy? And is our haste to develop this 'peaceful use' free of the historicist idea that we are moving into a new age—the 'atomic age'—and of similar nonsense?

its fruitfulness as a research programme). Practical usefulness or uselessness may be considered as important mainly because it is something like a test of truth—as it may often be in connection with a scientific theory.

But is it possible rationally to appraise or evaluate an irrefutable theory? What is the point of criticizing a theory rationally if we know from the start that it is neither refutable by pure reason, nor testable by experience?

My answer is this. If a metaphysical theory is a more or less isolated assertion, no more than the product of an intuition or an insight flung at us with an implied 'take it or leave it', then it may well be impossible to discuss it rationally. But the same would be true of a 'scientific' theory. Should anybody present us with the equations of classical mechanics without first explaining to us *what the problems are* which they are meant to solve, then we should not be able to discuss them rationally—no more than *The Book of Revelation.* Even if we are presented with Newton's arguments, we may be unable to discuss them unless we hear first about the problems of Galileo and Kepler and their solutions, and about Newton's own problem of how to unify these solutions by deriving them from a more general theory. In other words, any rational theory, no matter whether scientific or metaphysical, is rational only because it ties up with something else—because it is an attempt to solve certain problems; and it can be rationally discussed only *in relation to the problem situation* with which it is tied up. Any critical discussion of it will consist, in the main, in considering how well it solves its problems; how much better it does so than various competing theories; whether it does not create greater difficulties than those which it set out to dispel; whether the solution is simple; how fruitful it is in suggesting new problems and new solutions; and whether we cannot, perhaps, refute it by empirical tests.

This last method of discussing a theory is not, of course, applicable if the theory is metaphysical. But the other methods may well be applicable. This is why rational or critical discussion of some metaphysical theories is possible. (There may, of course, be other metaphysical theories which cannot be rationally discussed.)

It is hardly necessary to give examples of this method here, for this *Postscript* is full of them. I may mention as examples my critical

discussion of idealistic and positivistic epistemologies (in Volume I, Part 1); of probabilistic epistemologies (in Volume I, Part 2); and of deterministic metaphysics—as may be found in Hobbes, Hume, Kant, and Einstein (in Volume II).

I believe that my dream can be discussed in this way—especially by comparing it with competing views which it is meant to supplant. The comparison should be in terms of simplicity, coherence with certain other theories, unifying power, intuitive appeal and, above all, fruitfulness. Without wishing to commit myself to anything like pragmatism or instrumentalism, I think I should take the question of the fruitfulness of my programme as decisive. If it does not lead to new problems or, at least, to a new evaluation of some of the great old open problems, I should discard it: as a lovely dream (or so it seems to me)—lovely, yet not to be indulged in.

I do not wish here to give a list of the well-known open problems of the theory of matter (such as the problem of the electronic charge, or the derivation of the exclusion principle) and of general cosmology which (I dream) may perhaps be attacked one day in the light of the metaphysics of change which I have proposed here. But the problem of the large and still increasing number of elementary particles is one which has to be stressed because of its extreme urgency.

It was once the programme of modern atomic theory to explain the large numbers of atoms by *only two fundamental particles*—the electron and the proton; and its moment of greatest triumph occurred when it carried out this programme—a programme which now lies in ruins. The trouble started long ago, with the *ad hoc* hypothesis of the neutrino. It was introduced *ad hoc* in order to avoid defeat—a sound procedure in the case of a highly successful theory, as long as there is any hope of obtaining one day independent evidence for the *ad hoc* assumption. But if I am not mistaken, this hypothesis is still *ad hoc,* just as it was thirty years ago.[3] The neutrino, however, is small fry compared with all those other particles which have had to be introduced since, on compelling evidence—especially the various mesons; for they shattered the

[3][*(Added 1981) It was so when this book was being written.]

views of the structure of matter which had been the basis of quantum theory. Admittedly, these views were not part of the mathematical formalism; but they were part of physical theory none the less.

What is needed is a general theory of matter which explains the masses of all these particles, and their stability or instability, from general principles.[4] Einstein's demand for a unified theory is a necessary one, one which cannot be waved aside as a mere will o' the wisp, or as an illusion destroyed by quantum theory.[5] The theory of changing *fields of propensities* may offer a way towards unification.

Midway between atomic theory and cosmology proper there is the theory of the creation of matter due to Jordan, Gold, Hoyle, Bondi, and McCrea, which in intention at least is a scientific conjecture rather than a metaphysical theory (though it does not seem testable as yet). [*Not only has it become testable since, but it appears to have failed the test.] But assume a propensity theory of matter, as sketched in the foregoing section, and interpret space ('position' space, *topos*), following the suggestions of Leibniz and Einstein, as the field 'of the possible mutual relations of . . . bodies'[6]; that is to say, of the propensities of material particles to be positioned between other particles. (It is also, according to Dirac's theory, a field of states occupied by latent pairs of particles of matter, and capable of being polarized.) The expansion of the universe might then be interpreted as the expansion of this field of propensities. As this expansion creates new possibilities, and thus new propensities for matter to be present (some of which will be realized), it might thus explain the creation of new matter; for matter may be identified with the realization of these propensities. Thus an expanding universe might create matter merely as a function of its expansion (as perhaps suggested by the steady state theory).

In this connection, I want to draw attention to Kapp's conjecture that all material particles have a limited life-time, so that large accumulations of particles (stars, planets) would act like a 'sink' for matter (and thus, on the assumption just sketched, also for space); a

[4] A field theory which attempts to explain the masses of the particles (but not, to my knowledge, their lifetimes) has been proposed by M. Born and H. S. Green; *cp.* M. Born, *Rev. Mod. Phys.* 21, 1949, pp. 463–473.

[5] This was W. Heisenberg's thesis in *Dialectica* 2, 1948.

[6] A. Einstein, *Mein Weltbild*, p. 179; *cp. Essays in Science*, p. 14.

conjecture which, as Kapp suggests, might lead to a geometrical theory of gravitation similar to Einstein's.[7]

There are some cosmological problems which are closely connected with the idea of an indeterministic world. This idea implies, for example, the distinction between a *closed past* and an *open future,* and with it an objective direction of time. In our picture of the probabilistic or non-classical film strips each of which is attached to a time-slice of an indeterministic universe, we can say that even though the equations which determine the dynamic changes of propensities in the films may be symmetrical with respect to time, our theory is not. For our theory consists of more than these equations: it consists of these equations plus their interpretation in terms of propensities, and this interpretation is non-symmetrical with respect to time. The idea of propensity distinguishes the unrealized possibilities of the future from their realizations present and past.

This becomes clear in certain aspects of the quantum theory which I have discussed both in *The Logic of Scientific Discovery* and in this *Postscript. (Cp. L.Sc.D.*, section 73, note 5 and text, and section 16 of this volume of the *Postscript.)* As Heisenberg remarks, it is possible to calculate, with the help of certain measurements, for example of two positions of an electron, its spatio-temporal path between these two positions—*its momentum as well as its various positions*—with any desired degree of precision. Heisenberg added that it was 'a matter of personal belief whether such a calculation concerning the past history of the electron can be ascribed any physical reality or not'.[8] My own inquiry led to a different result: in my interpretation, these calculations were *necessary for testing* the Heisenberg formulae, interpreted as scatter relations, and therefore

[7]*Cp.* R. O. Kapp, *British Journal for the Philosophy of Science* **5**, 1955, pp. 331 *f.*, and **6**, 1955, pp. 177*ff.*, where further references are given. (To these should be added, P. Jordan, *Naturwiss.* **26**, 1938, p. 417; *Astr. Nachr.* **276**, 1948, p. 193.) I cannot help feeling that it would be preferable not to speak of a 'symmetry' between matter creation and matter extinction; in an expanding universe, at least, it appears that matter creation would have to keep the average density of matter constant, so that matter destruction, if any, would lead to an additional rate of matter creation, rather than to a corresponding or symmetrical rate of matter creation. Moreover the two rates, of creation and of destruction, would depend upon entirely different factors—the first upon changes of density, the second simply upon time.

[8]W. Heisenberg, *The Physical Principles of the Quantum Theory*, 1930, p. 20.

far from opposed to them. But the scatter relations do prevent similar calculations about the future. Thus we all agree that the future is open; and the fact, mentioned by Heisenberg, that we can calculate the past as determined, or closed, shows clearly that even in Heisenberg's interpretation there is an asymmetry here. In the propensity interpretation, the situation is perfectly clear. The future propensities are determined in every instant, *but only the* propensities. (This is why the scatter relations apply.) As these propensities realize themselves, in the form of particles with positions and momenta, the open possibilities become closed. Thus the calculability of the realizations, past and present, differs from that of the future. Although any one *given* time-slice does not allow us to determine the past any better than the future, sequences of *given* time-slices do determine the period between their first and last elements. *But any such sequence can only belong to the past.*

The idea of an indeterministic world bears upon, and offers a solution to, another problem which has been mentioned in this *Postscript*—Gödel's problem of the existence of closed world lines in certain cosmological solutions of Einstein's gravitational equations; that is to say, in rotating universes.[9] The history of a physical body moving on a closed world line would be repetitive, absolutely and infinitely repetitive. But this is possible only in a deterministic universe. The assumption of an indeterministic world therefore excludes this possibility; or rather, it will attribute to it the probability 0, whatever the initial conditions may be.

It is interesting from our point of view that the exclusion of this possibility by indeterminism is certainly not *ad hoc*.[10] But what seems even more interesting is the fact that, assuming Einstein's cosmological equations to be true (an assumption which seems to be compatible with our version of indeterminism), the exclusion of the solutions envisaged by Gödel amounts to the exclusion of certain ways for matter and its motion to be arranged in the world.[11] This

[9]K. Gödel, 'A Remark about the Relationship between Relativity Theory and Idealistic Philosophy', in *Albert Einstein: Philosopher-Scientist*, ed. P. A. Schilpp, pp. 555 *ff.*; and *Reviews of Modern Physics* 21, 1949, pp. 447–450. See also *The Open Universe: An Argument for Indeterminism*, Vol. II of this *Postscript*, note 2 to section 19.

[10]This fact seems to solve Gödel's problem. Gödel's suggestion that the solution lies in the rejection of realism and in the adoption of idealism, does not seem to me to offer any way out of his difficulty.

[11]K. Gödel, *op. cit.*, p. 562.

shows, surprisingly, that *a hypothesis like that of indeterminism can have structural consequences*—that the hypothesis of indeterminism may be able to exclude certain possible arrangements of matter and motion in the world. This fact is extremely interesting in view of what I have called 'Newton's Problem'. (See *Realism and the Aim of Science*, Volume I of the *Postscript*, Part 1, section 16.) It suggests that it may not be utterly impossible to make some headway in connection with certain aspects of this problem, even though it may well be insoluble in the main. But the exclusion, through the attribution of zero probability, of certain possible initial distributions of matter and motion has a further interesting consequence. It shows that in an indeterministic universe, there may be certain probabilistic principles of non-existence that exclude possibilities which from the point of view of a classical deterministic theory would appear typically contingent—possibilities of the character of initial conditions rather than of laws. (*Cp. L.Sc.D.*, Appendix *x.)

28. *Conclusion.*

The metaphysical programme of the propensity interpretation might be summed up, in the concise language of the Ionian cosmologists, by the statement: 'Everything is a propensity.' Or in the terminology of Aristotle we might say: 'To be is both to be the actualization of a prior propensity to become, and to be a propensity to become.' It is a view that unites aspects of all those metaphysical programmes listed in the first section of this epilogue (section 20), as will be seen from the following list.

1. As with Parmenides, the world is full—in the sense that the void, the vacuum, has a structure, and is itself a field of propensities which are real.

2. As with the atomists, the structure of matter is atomic, and the dualism of the full and the empty—or of matter and space or field—is, up to a point, preserved, as a distinction between the realization of a propensity, and the propensity to be realized. (See also position 10, below.)

3. As with Plato and Euclid, the emphasis upon geometry is preserved; and so is the geometrical cosmology: in this respect, non-Euclidean cosmology, as a cosmology, out-Euclids Euclid, for

geometry as such is used even to describe the distribution of matter in the world.

4. The Aristotelian view of inherent potentialities and their actualization is developed into a *relational* theory in which relational structures, instead of inhering in each material thing, may be characterized by potentialities.

5. The (Platonic) geometrical approach of the Renaissance is preserved, as well as Plato's hypothetical method, and its stress upon antecedent causes.

6. The theory of fluids (for example, heat) of the Cartesians and of Boyle is preserved in the form of the law of conservation of energy. Their action at vanishing distances is preserved in the form of the field theory.

7. The propensity theory may be described as a generalization of dynamism.

8. Central forces (which correspond to the Aristotelian inherent potentialities) give place, as with Faraday and Maxwell, to fields of potentialities of a relational character.

9. As in Einstein's and Schrödinger's programme, the dynamical laws of change of these fields of propensities are of a *prima facie* deterministic character (like the laws of a classical theory). Moreover, propensities—even those which for their geometrical representation need a multi-dimensional abstract space of possibilities—are treated as physical realities.[1] These two points, a

[1]Heisenberg (*Niels Bohr and the Development of Physics*, edited by W. Pauli, 1925, p. 24), tells us that probability waves in configuration space are not real, according to the orthodox Copenhagen interpretation which he calls 'the usual interpretation'. Thus the 'dualism of particle and wave' breaks down here: it holds only if the waves are in three-dimensional space. The passage (indeed the whole paper) is interesting, apart from other things, because of its queer suggestions, reminiscent of psycho-analysis, that the critics of Copenhagen are 'uninformed' (*cp. Realism and the Aim of Science*, Vol. I of *The Postscript*, Part I, section 18) about the esoteric teaching of the orthodoxy. 'Now Schrödinger's work', Heisenberg writes, 'first of all, contains some misunderstandings of the usual interpretation. He overlooks the fact that only the waves in configuration space . . . are probability waves in the usual interpretation, while the three-dimensional material waves or radiation waves are not. The latter . . . have just as much (and just as little) "objective reality" as particles; they have no direct connection with probability waves' I do not know whether Heisenberg intends to allow the Schrödinger waves for one and for two particles to be three-dimensional and thus 'real' (but not those for three particles), or whether he has only the 'second quantization' in mind; but in any case, Heisenberg's view seems to me to take three-dimensionality a shade

physical reality described by deterministic laws, and a reality to which the idea of field applies (so that the laws are partial differential equations), were Einstein's main ideas for a unified theory of the world. They are preserved.

10. The point of view of the orthodox interpretation of quantum theory, including its 'dualism' of field and particle, or of wave and particle, is re-interpreted and preserved in terms of the dualism of potentialities and their actualizations which in their turn are again potentialities. (This view transcends the older dualism, in a way often demanded by Einstein.) The probabilistic character of the theory, stressed especially by Born and Pauli, is also preserved.

Thus all the older programmes become approximations from the point of view of this metaphysical theory of propensities. All the programmes contribute some aspect or other to our metaphysical view.

In particular, the dispute between the last two positions on my list—between Einstein and Schrödinger on one side, and those who, like Pauli, uphold Born's essentially statistical interpretation of quantum theory on the other side—is resolved by the propensity interpretation: to Einstein and Schrödinger it offers a physical reality to which field equations of the classical deterministic type apply, and to Born it offers a probabilistic interpretation of these equations.

The propensity interpretation owes everything to Einstein. And yet I doubt very much whether he would have been prepared to accept it. What would have stood in the way was not only his determinism (cp. *The Open Universe: An Argument for Indeterminism,* Vol. II of the *Postscript,* section 26) but also his attitude towards probability, which was no doubt one of the results of his determinism. This attitude had been fundamentally subjective, and involved the usual assumption that statistical frequencies can be derived from a subjective theory. (There is a close resemblance at

too seriously (if he seriously means what he seems to say in this astonishing paper). After all, the multi-dimensional representations are only ways of putting things; and what they represent—the propensities and their laws—are just as real as what the three-dimensional waves represent. For it is clearly the same thing which is represented by the two methods. There is no difference of subject matter—not even of intended subject matter—but only between two more or less successful methods of representing it.

this point between Einstein's views and those of his opponents who, it seems, never succeeded in shaking off the eggshells of their determinist past.) When I say that the propensity interpretation owes everything to Einstein, I am thinking of his doctrine that the aim of physics is the description of physical reality in terms of field equations. At the same time, he was the first to operate with transition probabilities, that is to say, with *singular probabilities, pertaining to the single atom*, and proportional to the density of radiation.[2] At the end of the paper in which he developed this view, he stressed the task of establishing a link between Maxwell's wave theory and this new theory of his which, he said, 'leaves to chance both the instant and the direction of the elementary processes' of emission. Some time later, he connected ideas of Bose's and de Broglie's by suggesting that, just as it is possible to treat photons statistically like particles of a gas, so it must also be possible to associate with a gas a wave field.[3] In this context it is interesting to learn from a report of Bohr's that in 1920 Einstein described the situation in physics created by his photon theory (and by the refutations of Maxwell's theory implicit in the success of Bohr's model) by speaking of 'phantom fields that guide the photon';[4] the ghosts, or spectres, of the departed electromagnetic waves, were still something real (since they were still responsible for the spectra and especially for the intensity of spectral lines). But although Einstein practically anticipated the propensity theory, I do not believe that he would have liked it. (For other anticipations, see section 12 above.)

One striking advantage of the propensity interpretation is that it seems, if I may use Pauli's words, 'to open up hopes for a future development of different branches of science towards a greater unity'.[5] Pauli is here speaking of quantum theory in general, with its

[2]A. Einstein, *Physikalische Zeitschrift* 18, 1917, pp. 121–128.

[3]*Sitzungsberichte d. Preuss. Akad. d. Wissenschaften, Phys.–Math. Klasse*, 1925, pp. 3–14.

[4]Einstein's expression, '*Gespensterfelder*', is translated 'ghost waves' by Niels Bohr in 'Discussion with Einstein on Epistemological Problems in Atomic Physics', in *Albert Einstein: Philosopher-Scientist, op cit.*, p. 206.

[5]W. Pauli, 'Einstein's Contributions to Quantum Theory', in *Albert Einstein: Philosopher-Scientist*, pp. 157 f.

probabilistic laws; and no doubt, the different branches of science he has in mind are the biological sciences (see also note 4 to section 21), and ultimately the sciences of man. I think that his hopes are justified, and even more so in view of the propensity interpretation. Biologists have always worked with propensities—even when they were prevented by mechanistic prejudices from admitting it. The propensity interpretation of physics might help to remove these prejudices, and thereby encourage further mutual fertilization between the physical and the biological sciences. Take the case of chemical affinity. We may now say, in the light of its quantum theoretical explanation, that it is a physical propensity, and that the intuitive feeling that it was in many ways akin to certain biological propensities was justified: it is not merely an analogy we face here but, more likely, an intermediate, genuinely transitional case. The term 'propensity' itself is, of course, a biological or psychological term. (It is also used in economics, in precisely the same sense as, I suggest, it should be used in physics.)

A case of special interest is the old problem of the organization of living matter in individuals—individual plants, individual animals. The individual organisms of biology with their strange character of far-reaching autarky have often been compared to crystals; and indeed, they could be compared with physical systems that are endowed with strong propensities to retain their character as relatively autarkic systems—such as crystals, in the case of which these propensities are likely to be explicable in terms of resonance; or atomic nuclei whose astonishing propensities towards autarky are illustrated by the constancy of their rate of radioactivity upon which no environmental conditions short of bombardment by nucleons make any impression. We might call these peculiar propensities towards autarky, with their surprising independence of environmental conditions, *inherent propensities* of the system. They are relational, as all propensities are, of course; and yet, they do resemble Aristotle's inherent potentialities of a thing more than other physical or biological propensities. (This is no accident: Aristotle was a biologist.)

Ultimately, only within a framework of an indeterministic picture of the world can we ever hope to understand the phenomenon of voluntary movement in animals. (*Cf.* the footnote at the end of

section 25, above.) In biology many things happen, especially in animals, which, considered from a physical point of view, are most unexpected and improbable. That a pair of swallows, after flying long distances, repeatedly return to their old nesting place, is difficult to explain by physical laws, and the conservation laws do not seem to offer any help here. If memory can bring these things about, it must guide, somehow or other, the movements of these animals in a way similar to the way the pocket roulette-wheel guided the movements of our soldiers (*cp.* sections 10 and 25); except that the roulette-wheel is replaced by a kind of loadstone, so that the game of chance is replaced by an 'inherent propensity'. It hardly matters for our present purpose how these things function in detail. What matters is that they do function, and that they seem to function as if certain inherent propensities superimposed themselves (as in the Zeeman effect, say) upon certain other physical propensities of a more chance-like or equi-probabilistic character, thereby giving extra weight to certain possibilities: they impose, as it were, a systematic bias upon them. This seems to be the way in which so many improbable things happen in biological contexts. What we can now clearly see is that this kind of thing (I mean, the superposition of 'inherent propensities') already plays a role, in a rudimentary way, in classical physics (loaded dies, osmotic pressure, resonance); and we can therefore form an intuitive idea of how it may fit into our physical world, and yet transcend it, by superimposing upon it a hierachy of purposes—a hierarchy of systematic and increasingly purposeful biases.[6]

None of this is said in the spirit of a spiritualist's apology: man and his spirit need no apology. It is neither the conservation law of energy and momentum nor any other physical law, and not even a probability or a propensity, which has made him build the pyramids, or climb Everest; and he has reached still greater heights than this in science, in art, and in many other ways.

I have tried to present this metaphysical epilogue as what it is—a picture, a dream, rather than a testable theory. Science needs these pictures. They largely determine its problem situations. A new

[6][Compare Popper's discussions of biology in 'Of Clouds and Clocks', and 'Evolution and the Tree of Knowledge', in *Objective Knowledge*, 1972, pp. 206–84. Ed.]

picture, a new way of looking at things, a new interpretation may change the situation in science completely (as did Einstein's way of looking at the Lorentz transformations). But these pictures are not only much-needed tools of scientific discovery, or guides to it, they also help us to decide whether a scientific hypothesis is to be taken seriously; whether it is a potential discovery, and how its acceptance would affect the problem situation in science, and perhaps even the picture itself.

It is here, perhaps, that we may find a criterion of demarcation *within metaphysics*, between rationally worthless metaphysical systems, and metaphysical systems that are worth discussing, and worth thinking about. The proper aspiration of a metaphysician, I am inclined to say, is to gather all the true aspects of the world (and not merely its scientific aspects) into a unifying picture which may enlighten him and others, and which one day may become part of a still more comprehensive picture, a better picture, a truer picture. The criterion, then, will be fundamentally the same as in the sciences. Whether a picture is worth considering depends, I suggest, upon its capacity to provoke rational criticism, and to inspire attempts to supersede it by something better (rather than upon its capacity to create a fashion, to be supplanted presently by a new fashion, or upon claims to originality or finality). And this criterion, I believe, may also point to one of the characteristic differences between a work of science or of metaphysics and a work of art that aspires to be something that cannot, in its own way, be bettered.

INDICES

compiled by Stephen Kresge and Nancy Artis Sadoyama

References in italics indicate passages of special importance. A page number followed by '*t*' indicates that the term is discussed in the place referred to; '*n*' means 'footnote', and '*q*' (in the Index of Names) means 'quoted'. The reader is also advised to consult the Indices to the other volumes of the *Postscript*.

Index of Names

Index of Subjects

ity, 107; linked to lack of knowledge, 109, 115, 118. *See also* Entropy; Quantum theory, observer in; Statistical mechanics; Time reversibility

'KICKABILITY', and reality, 46, 72–73, 81, 84, 193
Knowledge, growth of, 33–34, *see also* Critical discussion; inherent limits to, xvii, 5, 7; lack of ('nescience'), linked to probabilistic physics, 4, 49, 75, 105–106, 108–109, 133, 145; alleged connection to lack of entropy, 109–117; role of, in quantum theoretical measurement, 35, *see* Quantum theory, observer in; *see also* Crucial experiment, extending the original EPR-argument

LACK OF KNOWLEDGE, 4, 7; link with statistics, 4, 49; Einstein's change of views concerning, 4. *See also* Knowledge, lack of
Law, of conservation, *see* Conservation laws; of large numbers, is no bridge between subjective beliefs and objective frequencies, 66–67, 106, *see also* Probability
Leibniz's dynamical theory of matter, 163, 166–171
Locality, principle of (principle of action at vanishing distances, 'Prinzip der Nahewirkung'), 19–21, 25–26; in clockwork theory, 163; in fields of forces theory, 164, 169; problem of, 16; and special relativity theory, 22n; support by Einstein, 21. *See also* Copenhagen interpretation; EPR-argument; Relativity
Logic, of the situation, 31t, 33
Lorentz transformations, xviii, 211

MANY WORLDS INTERPRETATION, of quantum theory, *see* Everett's 'many worlds' interpretation
Matter (body), creation of, 202; destructible, 164; dualism of, and field, *section* 26; electromagnetic theory of, xvii, 37–38, 192–194; essential properties of, according to Aristotle, 166, according to Hobbes, Descartes, and Boyle, 163, 166, according to Kant, 170–171

&n, according to Leibniz, 163, 166, 167, according to Mach, 172, according to Newton, 127, according to Poisson and Cauchy, 171; explanation of, 164; as formed space, 162–163; general theory of, 202; problem of, 30, 126–127, 165–172; as realization of propensities, 160, 202, *section* 26
Maxwell's demon, 114n
Meaning, doctrine of, in application to debate about quantum theory, 18n, 21, 39, 60–61, 102, 148, 150
Measurement, of a Doppler effect, 58; indirectness of, 41; of position and momentum, 17–19, 58–59, 60–64, 73–75, 142–143, 146–147, 155; possibility of, 5; in quantum mechanics, compared to classical physics, 35–41; retrodictive, of momentum, 146–147; retrodictive character, in tests of scatter relations, 60–64; verification of, 61. *See also* Crucial experiment, extending the original EPR-argument; Heisenberg's formulae; Quantum theory, observer in; Scatter relations
Mechanics, classical, statistical and particle, 47; equivalence of wave and matrix mechanics, 43; quantum mechanics, *see* Quantum theory; Statistical, *see* there
Meson, 13, 195
Metaphysical research programme, 31, 32t–34, *sections* 20 & 21; author's own, of the propensity interpretation and of physical propensities, *see* Propensity, metaphysical research programme of; Cartesian, 163, 166–167; clashes among, 173; cosmological speculations as, 31; create problems and determine solutions, *section* 20; criticizability of, 32–33, 172–173; examples of, 162–164; and 'normal science', 33; indispensable to science, 165; relation to Kuhn's paradigms, 31–36; relation to scientific theories, 32; and 'scientific research programmes', 32&n, 33–34; as speculative physics, 161–162, 165
Metaphysics, critical discussion of, 32, 172–173, 199–201; difference from science, 211; *sections* 27 & 28; task of, 211. *See also* Critical discussion; Demarcation criterion within, 211
Mind, human, reality of, xviii

221

ABOUT THE AUTHOR

Sir KARL (RAIMUND) POPPER was born in Vienna in 1902. He studied mathematics, physics, psychology, education, the history of music, and philosophy at the University of Vienna from 1918 to 1928. At the same time he worked as a cabinet maker's apprentice and as a teacher. In 1978, fifty years after he took his Ph.D. degree, the University of Vienna 'renewed' this degree in a solemn ceremony and gave him an honorary doctorate in the natural sciences.

In 1934, when still a schoolteacher in Vienna, he published *Logik der Forschung,* a book that became a classic after its translation into English. (*The Logic of Scientific Discovery* was published in 1959.) It has now been translated into many languages and, after 47 years, is still frequently reprinted. Popper, who lives in England, has lectured in Europe, in New Zealand, Australia, India, Japan, and since 1950, when he delivered the William James Lectures in Philosophy at Harvard University, often in America. Among his books are *The Open Society and Its Enemies* (for which he received the Lippincott Award of the American Political Science Association); *Conjectures and Refutations; The Poverty of Historicism; Objective Knowledge; Unended Quest.* Together with Sir John Eccles he published *The Self and Its Brain.*

He holds honorary degrees from the Universities of Chicago, Denver, Warwick, Canterbury (New Zealand), Salford, The City University (London), Vienna, Mannheim, Guelph (Canada), Frankfurt, Salzburg, Cambridge, and Oxford.

He is a Fellow of the Royal Society (London), and of the British Academy; a Foreign Honorary Member, American Academy of Arts and Sciences; Membre de l'Institut de France; Socio Straniero

227

dell'Accademia Nazionale dei Lincei; Membre de l'Académie Internationale de Philosophie des Sciences; Associate, Académie Royale de Belgique; Membre de l'Académie Européenne des Sciences, des Arts, et des Lettres; Hon. Member, the Royal Society of New Zealand; Membre d'Honneur, Académie Internationale d'Histoire des Sciences; Ehrenmitglied, Deutsche Akademie für Sprache und Dichtung.

He is also an honorary member, Harvard Chapter of Phi Beta Kappa; Ehrenmitglied, Allgemeine Gesellschaft für Philosophie in Deutschland; Hon. Fellow, London School of Economics and Political Science; and Hon. Fellow, Darwin College, Cambridge.

He was awarded the Prize of the City of Vienna for Moral and Mental Sciences; the Sonning Prize of the University of Copenhagen; the Dr Karl Renner Prize of the City of Vienna; and the Dr Leopold Lucas Prize of the University of Tübingen.

He received the Grand Decoration of Honour in Gold (Austria); the Gold Medal for Distinguished Service to Science of the American Museum of Natural History, New York; the Ehrenzeichen für Wissenschaft und Kunst (Austria); and the Order Pour le Mérite (German Federal Republic).

He was knighted in 1965.

ABOUT THE EDITOR

WILLIAM WARREN BARTLEY, III, a graduate of Harvard and the University of London, is a former student and colleague, and a long-time associate of Sir Karl Popper. He has been Lecturer in Logic at the London School of Economics, Lecturer in the History of the Philosophy of Science at the Warburg Institute, and S. A. Cook Bye-Fellow at Gonville and Caius College, Cambridge. Formerly Professor of Philosophy and of the History and Philosophy of Science at the University of Pittsburgh, he has since 1970 been Professor of Philosophy at California State University, Hayward, where, in 1979, he was named Outstanding Professor in the 19-campus California State University system.